"Whether the Earth is a single organism or not, Mr. Lovelock's illuminating arguments for his thesis should inspire respect and admiration for this clever planet."
—*The New York Times Book Review*

"Lovelock's hypothesis is backed by solid scientific research."
—*The Bloomsbury Review*

"As the prospect of global climatic change infuses the ecological sciences with new urgency, the Gaia hypothesis, whether true or false, appears to have significant influence on the way some scientists study the Earth and the questions they ask about it."
—*The New York Times*

"Lovelock's . . . books have struck home because he's sketched a hard-science framework on to an important emotional prejudice, and brought a whole bundle of disciplines to bear. . . . *The Ages of Gaia* [refines] his theory of the Earth's prehistory."   —*Arena*

"Instead of seeking a safe niche in a scientific subspecialty, the maverick chemist-biologist-inventor has spent two decades working on an all-encompassing theory of evolution—one that will explain not only how the camel got his hump and the peacock its plume but how the Earth came by its unique climate, chemistry, and atmosphere."   —*Newsweek*

"By turns scholarly, combative, sentimental: Lovelock's argument persuades through its passion as well as its logic. A worthy tribute to a mighty Lady."   —*Kirkus Reviews*

BANTAM NEW AGE BOOKS

This important imprint includes books in a variety of fields and disciplines and deals with the search for meaning, growth and change.

Ask your bookseller for the books you have missed.

The Commonwealth Fund Book Program
gratefully acknowledges the assistance of
Memorial Sloan-Kettering Cancer Center
in the administration of the Program.

The Ages of
GAIA

A Biography of
Our Living Earth
BY
James Lovelock

A volume of

THE COMMONWEALTH FUND

BOOK PROGRAM

under the editorship of Lewis Thomas, M.D.

BANTAM BOOKS
NEW YORK · TORONTO · LONDON · SYDNEY · AUCKLAND

THE AGES OF GAIA

*A Bantam Book / published by arrangement with
W. W. Norton and Company, Inc.*

PRINTING HISTORY
W. W. Norton edition published September 1988

*Bantam New Age and the accompanying figure design as well as the statement
"the search for meaning, growth and change" are trademarks of Bantam Books, a
division of Bantam Doubleday Dell Publishing Group, Inc.*
Bantam edition / April 1990

Library of Congress Cataloging-in-Publication Data
Lovelock, J. E.
  The ages of Gaia : a biography of our living earth / by James
Lovelock,
    p.   cm.—(The Commonwealth Fund Book Program)
  Reprint. Originally published: New York : Norton, c1988.
  Includes bibliographical references.
  ISBN 0-553-34816-7
  1. Biology—Philosophy. 2. Biosphere. 3. Life (Biology)
I. Title.   II. Series: Commonwealth Fund Book Program (Series)
QH331.L688 1990
575'.01—dc20   89-17577
                                                                 CIP

Published simultaneously in the United States and Canada

Bantam Books are published by Bantam Books, a division of Bantam Doubleday
Dell Publishing Group, Inc. Its trademark, consisting of the words "Bantam
Books" and the portrayal of a rooster, is Registered in U.S. Patent and Trademark
Office and in other countries. Marca Registrada. Bantam Books, 666 Fifth
Avenue, New York, New York 10103.

Contents

Foreword

Most working scientists have an awareness and respect for the history of the fields in which they labor, but what they generally have in mind is a series of endeavors strung through the volumes of their specialized journals that are still held in the library stacks—not at all the much longer stretch of time and work that professional scholars would require for a proper history of science.

It is not that researchers have short memories, but that they learn and retain only the events that set their fields atremble in the first place. And for most of science these days, perhaps all of it, the great changes that launched this century's vast transformation of human knowledge began within this century, or at least seemed to. The modern postdoctoral student in a laboratory engaged in molecular biology, for instance, feels no dependence on generations of forebears more than 20 years back. The contemporary physicists may track their ideas back almost a century, to the beginnings of quantum theory, but it is the concepts emerging in only the past decade that are regarded as the real history. The cosmologists are out on totally new ground, looking in amazement at strange, unanticipated kinds of space and time, making educated guesses at phenomena far beyond the suburban solar system or the local galaxy, even speculating about universes bubbling out at the boundaries of this one.

We are, quite literally, in a new world, a much more peculiar place than it seemed a few centuries back, harder to make sense of, riskier to speculate about, and alive with information which is becoming more accessible and bewildering at the same time. It sometimes seems that there is not just more to be learned, there is *everything* to be learned.

This is far from the general public view of the matter, as reflected in the science sections of newspapers and newsmagazines. The nonscientific layman tends to take technology to be so closely linked to science as to be the center of the enterprise. The progress of science and that of technology seem to be all of a piece—machines, electronics, computer chips, Mars landings, nonbiodegradable plastics, the ozone hole, the bomb, all the rest of what now looks like twentieth-century culture.

What is not so clearly seen is the newness of the scientific information itself, the strangeness, and, where meaning is to be discerned, the meaning. There is a great difference between the intellectual product of modern science and the various technologies that are sometimes (nothing like as frequently as the public might guess) derived from that product.

The books in this series, sponsored by The Commonwealth Fund, represent an attempt to clarify this distinction, as well as to provide a closer look at what goes on in the minds of scientists as they go about their work.

This book by James Lovelock describes a set of observations about the life of our planet which may, one day, be recognized as one of the major discontinuities in human thought. If Lovelock turns out to be as right in his view of things as I believe he is, we will be viewing the Earth as a coherent system of life, self-regulating, self-changing, a sort of immense organism. This is not likely, in my opinion, to lead directly or indirectly to any specific piece of new technology to be put to use, although it may very well begin to influence, in new and gentler ways, the other sorts of technology we might be selecting for use in the future.

The selection of this book, and of others in The Commonwealth Fund Book Program, has been the responsibility of an Advisory Committee consisting of: Alexander G. Bearn, M.D.; Donald S. Fredrickson, M.D.; Lynn Margulis, Ph.D.; Maclyn McCarty, M.D.; Lady Jean Medawar; Berton Roueché; Frederick Seitz, Ph.D.; and Otto Westphal, M.D. The publisher is represented by Edwin Barber, senior vice president, W. W. Norton & Company. Antonina W. Bouis serves as managing editor of the series, and Stephanie Hemmert as secretary. Margaret Mahoney, president of The Commonwealth Fund, has actively supported the work of the Advisory Committee at every turn.

LEWIS THOMAS, M.D., Editor,
The Commonwealth Fund Book Program

Preface

I am writing in a room added to what was once a water mill that drew power from the River Carey as it flowed on to join the Tamar and the sea. Coombe Mill is still a work place, but now a laboratory, den, and meeting place where I spend much of my time. The room looks out onto the river valley with its small fields and hedgerows typical of the Devonshire country scene.

The description of the place where this book was written is relevant to its understanding. I work here and it is my home. There is no other way to work on an unconventional topic such as Gaia. The researches and expeditions to discover Gaia have occupied nearly twenty years, and have been paid for from the income I receive for the invention and development of scientific instruments. I gratefully acknowledge the generosity of Helen Lovelock for letting me use the greater part of our joint income this way and also the faithful and consistent role of the Hewlett Packard Company, who have been the best of customers for my inventions, and truly have made the research possible.

Science, unlike other intellectual activities, is almost never done at home. Modern science has become as professional as the advertising industry. And, like that industry, it relies on an expensive and exquisitely refined technique. There is no place for the amateur in modern science, yet, as is often the way with professions, science more often applies its expertise to the trivial than to the numinous. Where science differs from the media is in its lack of a partnership with independent individuals. Painters, poets, and composers easily move from their own worlds into advertising and back again, and both worlds are enriched. But where are the independent scientists?

You may think of the academic scientist as the analogue of the independent artist. In fact, nearly all scientists are employed by some large organization, such as a governmental department, a university, or a multinational company. Only rarely are they free to express their science as a personal view. They may think that they are free, but in reality they are, nearly all of them, employees; they have traded freedom of thought for good working conditions, a steady income, tenure, and a pension. They are also constrained by an army of bureaucratic forces, from the funding agencies to the health and safety organizations. Scientists are also constrained by the tribal rules of the discipline to which they belong. A physicist would find it hard to do chemistry and a biologist would find physics well-nigh impossible to do. To cap it all, in recent years the "purity" of science is ever more closely guarded by a self-imposed inquisition called the peer review. This well-meaning but narrow-minded nanny of an institution ensures that scientists work according to conventional wisdom and not as curiosity or inspiration moves them. Lacking freedom they are in danger of succumbing to a finicky gentility or of becoming, like medieval theologians, the creatures of dogma.

As a university scientist I would have found it nearly impossible to do full-time research on the Earth as a living planet. To start with, there would be no funds approved for so speculative a research. If I had persisted and worked in my lunch hour or spare time, it would not have been long before I received a summons from the lab director. In his office I would have been warned of the dangers to my career of persisting in so unfashionable a research topic. If this did not work and obstinately I persisted, I would have been summoned a second time and warned that my work endangered the reputation of the department, and the director's own career.

I wrote the first Gaia book so that a dictionary was the only aid needed and I have tried to write this way in the present book. I am puzzled by the response of some of my scientific colleagues who take me to task for presenting science this way. Things have taken a strange turn in recent years; almost the full circle from Galileo's famous struggle with the theological establishment. It is the scientific establishment that makes itself esoteric and is the scourge of heresy.

It was not always like this. You may well ask, Whatever became of those colorful romantic figures, the mad professors, the Drs. Who? Scientists who seemed to be free to range over all of the disciplines of science without let or hindrance? They still exist, and in some ways I am writing as a member of their rare and endangered species.

More seriously, I have had to become a radical scientist also because the scientific community is reluctant to accept new theories as fact, and rightly so. It was nearly 150 years before the notion that heat is a measure of the speed of molecules became a fact of science, and 40 years before plate tectonics was accepted by the scientific community.

Now perhaps you see why I work at home supporting myself and my family by whatever means come to hand. It is no penance, rather a delightful way of life that painters and novelists have always known about. Fellow scientists join me, you have nothing to lose but your grants.

The main part of this book, chapters 2 to 6, is about a new theory of evolution, one that does not deny Darwin's great vision but adds to it by observing that the evolution of the species of organisms is not independent of the evolution of their material environment. Indeed the species and their environment are tightly coupled and evolve as a single system. What I shall be describing is the evolution of the largest living organism, Gaia.

My first thoughts about Gaia came when I was working in Norman Horowitz's biosciences division of the Jet Propulsion Laboratory, where we were concerned with the detection of life on other planets. These preliminary ideas were expressed briefly in the proceedings of a meeting held by the American Astronautical Society in 1968 and more definitely in a letter to *Atmospheric Environment* in 1971. But it was not until two years later, following an intense and rewarding collaboration with the biologist Lynn Margulis, that the skeleton Gaia hypothesis grew flesh and came alive. The first papers were published in the journals *Tellus* and *Icarus,* where the editors were sympathetic and were prepared to see our ideas discussed.

Lynn Margulis is the staunchest and best of my colleagues. I am also fortunate in that she is unique among biologists in her broad understanding of the living world and its environment. At a time when biology has divided itself into some thirty or more narrow specialties proud in their ignorance of other sciences, even of other biological disciplines, it needed someone with Lynn's rare breadth of vision to establish a biological context for Gaia.

Sometimes, when confronted with excesses of sentiment about life on Earth, I follow Lynn's lead and take on the role of shop steward, the trade union representative, of microorganisms and the lesser under-represented forms of life. They have worked to keep this planet fit for life for 3.5 billion years. The cuddly animals, the wildflowers, and the people are to be revered, but they would be as nothing were it not for the vast infrastructure of the microbes.

It would be difficult after spending nearly twenty years developing a theory of the Earth as a living organism—where the evolution of the species and their material environment are tightly coupled but still evolve by natural selection—to avoid

capturing views about the problems of pollution and the degradation of the natural environment by humans.

Gaia theory forces a planetary perspective. It is the health
of the planet that matters, not that of some individual species
of organisms. This is where Gaia and the environmental movements, which are concerned first with the health of people,
part company. The health of the Earth is most threatened by
major changes in natural ecosystems. Agriculture, forestry, and
to a lesser extent fishing are seen as the most serious sources
of this kind of damage with the inexorable increase of the greenhouse gases, carbon dioxide, methane, and several others coming
next. Geophysiologists do not ignore the depletion of the ozone
layer in the stratosphere with its concomitant risk of increased
irradiation with short-wave ultraviolet, or the problem of acid
rain. These are seen as real and potentially serious hazards but
mainly to the people and ecosystems of the First World—from
a Gaian perspective, a region that is clearly expendable. It was
buried beneath glaciers, or was icy tundra, only 10,000 years
ago. As for what seems to be the greatest concern, nuclear
radiation, fearful though it is to individual humans is to Gaia
a minor affair. It may seem to many readers that I am mocking
those environmental scientists whose life work is concerned
with these threats to human life. This is not my intention. I
wish only to speak out for Gaia because there are so few
who do, compared with the multitudes who speak for the
people.

Because of this difference in emphasis, a concern for the planet
rather than for ourselves, I came to realise that there might be
the need for a new profession, that of planetary medicine. I
am indebted to the historian Donald McIntyre for writing to
tell me that it was James Hutton who first introduced the idea
of planetary physiology in the eighteenth century. Hutton was
a physician as well as a geologist. Physiology was the first science
of medicine, and one of the aims of this book is to establish
"geophysiology" as a basis for planetary medicine. At this early
stage in our understanding of the Earth as a physiological entity,
we need general practitioners, not specialists. We are like physi-

cians before the use of antibiotics; even in the 1930s they could offer little other than symptomatic relief to patients suffering from infection. Now the major causes of death in the early part of this century—tuberculosis, diphtheria, whooping cough, pneumonia—have vastly declined, and physicians are mostly concerned with the diseases of degeneration—cardiovascular and neoplastic disease. It is true that the appearance of the HIV virus has shaken the confidence we had in medicine to cure all ills, but still we have advanced far beyond the helplessness of the days before the 1940s.

We are in the same condition now with respect to the health of the Earth as were the early physicians. Specialties, like biogeochemistry, theoretical ecology, and evolutionary biology, all exist, but they have no more to offer the concerned environmental physician or the patient than could the analogous science of biochemistry and microbiology in the nineteenth century.

As part of their graduation, physicians must take the Hippocratic Oath. It includes the injunction to do nothing that would harm the patient. A similar oath is needed for putative planetary doctors if they are to avoid iatrogenic error: an oath to prevent the overzealous from applying a cure that would do more harm than good. Consider, for example, an industrial disaster that contaminated a whole geographic region with easily measurable levels of some carcinogenic agent, one that posed a calculable risk to the whole population of the region. Should all the food crops and livestock of the region be destroyed to prevent the risk attendant upon their consumption? Should nature, instead, just take its course? Or should we aim for some less stark choice in between? A recent disaster illustrates how, in the absence of the voice of the planetary physician, treatment may be applied with consequences more severe than those of the poison. I refer to the tragedy of Swedish Lapland in the wake of the Chernobyl accident, where thousands of reindeer, the food prey of the Lapps, were destroyed, because it was thought they were too radioactive to eat. Was it justifiable to inflict this brutal treatment for mild radioactive poisoning on a fragile culture and its dependent ecosystem? Or were the consequences of the "cure" worse than the remote and theoretical risk of cancer in a small proportion of its inhabitants?

In addition to a chapter on these environmental affairs, the last part of the book is concerned with some speculations about establishing a geophysiological system on Mars. The first Gaia book also stirred an interest in the religious aspects of Gaia, so in another chapter I have tried to answer some of the difficult questions that were raised. In this unfamiliar territory I have benefited from the strong moral support of the Lindisfarne Fellowship and especially from its founders, William Irwin Thompson and James Morton, and from the friendship of its other members, like Mary Catherine Bateson, John and Nancy Todd, and Stewart Brand, who was for many years the editor of CoEvolution Quarterly.

From the days when I first started writing and thinking about Gaia I have been constantly reminded how often the same general idea has been posed before, I have felt a special empathy with the writings of the ecologist Eugene Odum. If I unintentionally offend prior "geophysiologists" by failing to give credit to their writings, I ask their forgiveness. I know that there must be numerous other thinkers, like the Bulgarian philosopher, Stephen Zivadin, who have said much of it before me and have been ignored.

I have been fortunate in the friends who have read and criticized the chapters of the book as it was written. Peter Fellgett, Gail Fleischaker, Robert Garrels, Peter Liss, Andrew Lovelock, Lynn Margulis, Euan Nisbet, Andrew Watson, Peter Westbroek, and Michael Whitfield, all have unstintingly and thoughtfully given their advice on the science. I am equally thankful to my friends who criticised the book in terms of its readability: Alex and Joyce Andrew, Stewart Brand, Peter Bunyard, Christine Curthoys, Jane Gifford, Edward Goldsmith, Adam Hart-Davis, Mary McGowan, and Elizabeth Sachtouris. Since 1982 the United Nations University, through its program officer, Walter Shearer, has provided moral and material support especially for the notion of planetary medicine.

Left to myself I tend to write blocks of text that, like the pattern of a mosaic, make sense only from a distant view. I greatly valued the friendly skill with which Jackie Wilson re-arranged my words as she edited the manuscript and made it readable.

The Commonwealth Fund Book Program, by their generous support, gave me the opportunity to set aside the time needed to develop the ideas of the book and to write. I am especially grateful to Lewis Thomas and to the two editors Helene Friedman and Antonina Bouis for the warmth of their encouragement and moral support.

But this book could never have been written without the sustenance and love that Helen and John Lovelock so freely gave.

Viewed from the distance of the moon,
the astonishing thing about the earth,
catching the breath, is that it is alive. The
photographs show the dry, pounded surface
of the moon in the foreground, dead as
an old bone. Aloft, floating free beneath
the moist, gleaming membrane of bright
blue sky, is the rising earth, the only
exuberant thing in this part of the cosmos.
If you could look long enough, you would
see the swirling of the great drifts of white
cloud, covering and uncovering the half-
hidden masses of land. If you had been
looking a very long, geologic time, you
could have seen the continents themselves
in motion, drifting apart on their crustal
plates, held aloft by the fire beneath. It
has the organized, self-contained look of
a live creature, full of information,
marvelously skilled in handling the sun.

LEWIS THOMAS, *The Lives of a Cell*

THE AGES OF GAIA

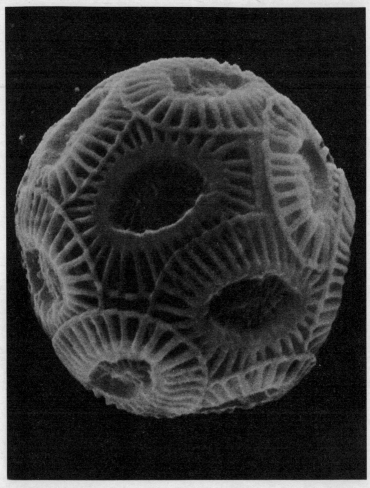

Emiliana huxleyii, known by her friends as Emily, is one of the more important members of the biota. Blooms of these phytoplankton cover large areas of ocean; their presence powerfully affects the environment through their capacity to facilitate the removal of carbon dioxide from the air and their production of dimethyl sulfide (which acts to nucleate clouds over the oceans).

1

Introductory

All through my boyhood I had a profound conviction that I was no good, that I was wasting my time, wrecking my talents, behaving with monstrous folly and wickedness and ingratitude—and all this, it seemed, was inescapable, because I lived among laws which were absolute, like the law of gravity, but which it was not possible for me to keep.

GEORGE ORWELL, *A Collection of Essays*

Of all the prizes that come from surviving more than fifty years the best is the freedom to be eccentric. What a joy to be able to explore the physical and mental bounds of existence in safety and comfort, without bothering whether I look or sound foolish. The young usually find the constraints of convention too heavy to escape, except as part of a cult. The middle-aged have no time to spare from the conservative business of living. Only the old can happily make fools of themselves.

The idea that the Earth is alive is at the outer bounds of scientific credibility. I started to think and then to write about it in my early fifties. I was just old enough to be radical without the taint of senile delinquency. My contemporary and fellow villager, the novelist William Golding, suggested that anything alive deserves a name—what better for a living planet than Gaia, the name the Greeks used for the Earth Goddess?

The concept that the Earth is actively maintained and regu-

3

lated by life on the surface had its origins in the search for life on Mars. It all started one morning in the spring of 1961 when the postman brought a letter that was for me almost as full of promise and excitement as a first love letter. It was an invitation from NASA to be an experimenter on its first lunar instrument mission. The letter was from Abe Silverstein, director of the NASA space flight operations. I can still recall the joyous and lasting incredulity.

Space is only a hundred miles away and is now a common place. But 1961 was only four years after the Soviet Union had launched the first artificial satellite, *Sputnik*. I listened to it bleeping its simple message that showed we could escape from Earth. Only six years earlier a distinguished astronomer said, when asked what he thought of the possibility of an artificial satellite, "Utter bunk." To receive an official invitation to join in the first exploration of the Moon was a legitimization and recognition of my private world of fantasy. My childhood reading had moved on that well-known path from *Grimm's Fairy Tales* through *Alice's Adventures in Wonderland* to Jules Verne and H. G. Wells. I had often said in jest that it was the task of scientists to reduce science fiction to practice. Someone had listened and called my bluff.

My first encounter with the space science of NASA was to visit that open-plan cathedral of science and engineering, the Jet Propulsion Laboratory, just outside the suburb of Pasadena in California. Soon after I began work with NASA on the lunar probe, I was moved to the even more exciting job of designing sensitive instruments that would analyze the surfaces and atmospheres of the planets. My background, though, was biology and medicine, and I grew curious about the experiments to detect life on other planets. I expected to find biologists engaged in designing experiments and instruments as wonderful and exquisitely constructed as the spacecraft themselves. The reality was a disappointment that marked the end of my euphoria. I felt that their experiments had little chance of finding life on Mars, even if the planet were swarming with it.

When a large organization is faced with a difficult problem

the standard procedure is to hire some experts, and this is what NASA did. This approach is fine if you need to design a better rocket engine. But if the goal is to detect life on Mars, there are no such experts. There were no Professors of Life on Mars, so NASA had to settle for experts of life on Earth. These tended to be biologists familiar with the limited range of living things that they work with in their Earth-bound laboratories. There was no reason to suppose that such life forms would exist on Mars, even if life there were plentiful.

From the beginning to the end, the Martian life-detection experiments had a marked air of unreality. Let me illustrate this with a fable. Dr. X, an eminent biologist, showed me his Martian life detector; a cubical cage of stainless steel, beautifully constructed, with sides about one centimeter long. When I asked him how it worked, he replied, "It's a flea trap. Fleas are attracted to the bait inside, hop in, and cannot escape." I then asked how he could be sure that there are fleas on Mars; his response was, "Mars is the greatest desert in the Solar System—a planet full of desert. Wherever there are deserts there are camels, and there is no animal with as many fleas as a camel. On Mars my detector will not fail to find life." I think that the other scientists at the Jet Propulsion Laboratory tolerated me as a devil's advocate. They were under great pressure to get on with the job, and so had little time to think about what the job really was. They viewed my questions about Martian fleas with amused skepticism.

I was sure that there was a better way. At that time Dian Hitchcock, a philosopher, visited the Jet Propulsion Laboratory, where she was employed by NASA to assess the logical consistency of the experiments. Together we decided that the most certain way to detect life on planets was to analyze their atmospheres. We published two papers suggesting that life on a planet would be obliged to use the atmosphere and oceans as conveyors of raw materials and depositories for the products of its metabolism. This would change the chemical composition of the atmosphere so as to render it recognizably different from the atmosphere of a lifeless planet. Even on Earth the Viking lander might have failed to find life had it landed on the antarctic ice. By contrast a full atmospheric analysis, which the Viking was not equipped to do, would have provided a clear answer; indeed, even in the 1960s, analyses of the Martian atmosphere were available from telescopes that used infrared instead of visible light to look at Mars. They revealed an atmosphere that was dominated by carbon dioxide and not far from the state of chemical equilibrium. The gases in the Earth's atmosphere, on the other hand, are in a persistent state of disequilibrium. This strongly suggested to us that Mars was lifeless.

This conclusion was not popular with our NASA sponsors. They badly needed reasons to support the cost of a Mars expedition, and what goal could be more enticing than the discovery of life there? A certain Senator Proxmire, staunch guardian of the public purse, might have been interested to learn that NASA was pressing on with a Martian landing, at great expense, even when scientists within the organization had said there could be no life there to find. He might have been outraged had he discovered that as part of our research, supported by NASA funds, Hitchcock and I had turned an imaginary telescope on our own planet to show that the Earth bore life in abundance.

During those exciting days we often argued about the life that might be on Mars and about the extent of its cover of the surface. In the late 1960s, NASA sent its Mariner spacecraft to view the surface from orbit around the planet. Their view showed Mars, like the Moon, to be extensively cratered, and tended to confirm the dismal prediction that Dian Hitchcock and I had made from a study of its atmospheric composition; that it was probably lifeless. I recall a gentle discussion with Carl Sagan, who thought it might still be possible that life existed in oases where local conditions would be more favorable. Long before Viking set course from Earth I felt intuitively that life could not exist on a planet sparsely; it could not hang on in a few oases, except at the beginning or at the end of its tenure. As Gaia theory developed, this intuition grew; now I view it as a fact.

There was much argument about the need to sterilize the

spacecraft before sending them to Mars. I could never understand why it should be thought so bad to run the small risk of accidentally seeding Mars with life; it might even be the only chance we had of passing life on to another planet. Sometimes the argument was fierce and macho; full of adolescent masculinity. In any event, feeling as I did—that Mars was dead—the image of rape, sometimes used, could not be sustained; at worst the act would be only the dismal lonely aberration of necrophilia. More seriously, as an instrument designer I knew that the act of sterilization made all but impossible the already superhuman task of building the Vikings and threatened the integrity of their exquisitely engineered internal homeostasis.

To this day I appreciate the toleration and generosity of my colleagues at the Jet Propulsion Laboratory and in NASA, especially the personal kindness of Norman Horowitz, who was then head of the team of space biologists. In spite of the "bad news" I had brought, they continued to support my researches until the Viking missions to Mars were ready to go. The soft landing on Mars in 1975 of these two intricate and almost humanly intelligent robots was successful. Their mission was to find life on Mars, but the messages they returned as radio signals to the Earth returned only the chill news of its absence. Mars, except during day in the summer, was a place of pitiless frigidity, and implacably hostile to the warm wet life of Earth. The two Vikings now sit there brooding silently, no longer allowed to report the news from Mars, hunched against their final destruction by the wind with its burden of abrasive dust and corrosive acid. We have accepted the barrenness of the Solar System. The quest for life elsewhere is no longer an urgent scientific goal, but the confirmation by the Vikings of the utter sterility of Mars has hung as a dark contrasting backcloth for new models and images of the Earth. We now understand that our planet differs greatly from her two dead siblings, Mars and Venus.

That, then, is how the Gaia hypothesis started. We looked at the Earth in our imagination, and therefore with fresh eyes, and found many things, including the radiation from the Earth of an infrared signal characteristic of the anomalous chemical composition of its atmosphere. This unceasing song of life is audible to anyone with a receiver, even from outside the Solar System. I will try to show in the chapters that follow that unless life takes charge of its planet, and occupies it extensively, the conditions of its tenancy are not met. Planetary life must be able to regulate its climate and chemical state. Part-time or incomplete occupancy or mere occasional visits will not be enough to overcome the ineluctable forces that drive the chemical and physical evolution of a planet. The imaginary exercise of seeding Mars with life, or even of bringing Mars to life, is discussed in chapter 8. It is about the effort needed to bring Mars to a state fit for life and to maintain it in that state until life has taken charge. It illustrates the awesome extent to which the greater part of our own environment on Earth is always perfect and comfortable for life. The energy of sunlight is so well shared that regulation is, effectively, free of charge.

The Gaia hypothesis supposes the Earth to be alive, and considers what evidence there is for and against the supposition. I first put it before my fellow scientists in 1972 as a note with the title "Gaia as Seen Through the Atmosphere." It was brief, taking only one page of the journal *Atmospheric Environment*. The evidence was mostly drawn from the atmospheric composition of the Earth and its state of chemical disequilibrium. This evidence is reviewed in table 1.1 in comparison with modern knowledge of the compositions of the atmospheres of Mars and Venus, and with a guess at the atmosphere the Earth might have now, had it never known life. After long and intense discussions, Lynn Margulis and I produced more detailed yet concise statements in the journals *Tellus* and *Icarus*. Then in 1979, Oxford University Press published my book *Gaia: A New Look at Life on Earth*, which collected all our ideas up to that point. I began to write that book in 1976, when NASA's Viking spacecraft were about to land on Mars. I used their presence there as planetary explorers to set the scene for the discovery of Gaia, the largest living organism in the Solar System.

Ten years on and it is time to write again; this time about

Table 1.1 PLANETARY ATMOSPHERES: THEIR COMPOSITION

		Planet		
GAS	VENUS	EARTH WITHOUT LIFE	MARS	EARTH AS IT IS
Carbon dioxide	96.5%	98%	95%	0.03%
Nitrogen	3.5%	1.9%	2.7%	79%
Oxygen	trace	0.0	0.13%	21%
Argon	70 ppm	0.1%	1.6%	1%
Methane	0.0	0.0	0.0	1.7 ppm
Surface temperatures °C	459	240 to 340	−53	13
Total pressure, bars	90	60	0.0064	1.0

getting to know Gaia and discovering what kind of life she is. The simplest way to explore Gaia is on foot. How else can you so easily be part of her ambience? How else can you reach out to her with all your senses? I was delighted a few years ago, therefore, to read of another man who enjoyed walking in the countryside and who also believed the Earth to be alive.

Yevgraf Maksimovich Korolenko lived over 100 years ago in Kharkov in the Ukraine. He was an independent scientist and philosopher. He too was in his sixties when he began to express and discuss ideas far too radical for the merely middle-aged. Korolenko was a learned man; although self-educated, he was familiar with the works of the great natural scientists of his time. He did not recognize any authority, philosophical, religious, or scientific, but tried to discover answers for himself. One of those with whom he shared his country walks and his radical ideas was his young cousin, Vladimir Vernadsky. Vernadsky, who was to become an outstanding Soviet scientist, was deeply impressed by the old man's assertion that "The Earth is an organism." But to Vernadsky's biographer, R. K. Balandin, this "is another of Korolenko's aphorisms. It is doubtful that young Vladimir Vernadsky should have remembered this aphorism half a century later. Nevertheless, Korolenko's naive analogy

of the Earth as a living organism could not but excite the imagination of his young friend."

The idea that the Earth is alive is probably as old as humankind. But the first public expression of it as a fact of science
was by a Scottish scientist, James Hutton. In 1785 he said, at
a meeting of the Royal Society of Edinburgh, that the Earth
was a superorganism and that its proper study should be physiology. He went on to compare the cycling of the nutritious elements
in the soil, and the movement of water from the oceans to the
land, with the circulation of the blood. James Hutton is rightly
remembered as the father of geology, but his idea of a living
Earth was forgotten, or denied, in the intense reductionism of
the nineteenth century—except in the minds of isolated philosophers like Korolenko.

Today, we all use the word "biosphere" rarely recognizing
that it was Eduard Suess who in 1875 first used the term, in
passing, when describing his work on the geological structure
of the Alps. Vernadsky developed the concept, and from 1911
used its modern meaning. Vernadsky said: "The biosphere is
the envelope of life, i.e. the area of living matter . . . the biosphere can be regarded as the area of the Earth's crust occupied
by transformers which convert cosmic radiations into effective
terrestrial energy: electrical, chemical, mechanical, thermal, etc."

When I first formulated the Gaia hypothesis, I was entirely
ignorant of the related ideas of these earlier scientists, especially
Hutton, Korolenko, and Vernadsky. I was also unaware of similar
ideas expressed in recent years by many scientists, such as Alfred
Lotka, the founder of population biology, Arthur Redfield, an
ocean chemist, and J. Z. Young, a biologist. I acknowledged
only the inspiration of G. E. Hutchinson, a distinguished limnologist at Yale University, and of Lars Sillén, a Swedish geochemist.
But I was not alone in this ignorance; in the vigorous objections
to or support for Gaia made by colleagues in all sciences, none
observed that what was said followed naturally from Vernadsky's view of the world. Even as late as 1983, the monumental
Earth's Earliest Biosphere, edited by geologist J. W. Schopf and
including contributions from twenty of the most distinguished

American and European Earth scientists, made no mention of either Hutton or Vernadsky.

The all-too-common deafness of English speakers to any other language kept from our common knowledge the everyday science of the Russian-speaking world. It would be easy to attribute the lack of recognition of Vernadsky's contributions to the present political divisions, but, although this may play some part, I think that it is a small one compared with the malign effects of the nineteenth-century separation of science into neat compartments where specialists and experts could ply their professions in complacency. How many physicists are proud of their ignorance of what they call the "soft sciences"? How many biochemists can name the wildflowers of their countryside? In such a climate of opinion it is not surprising that Vernadsky's biographer found Korolenko's statement, "The Earth is a living organism," to be naive. Most scientists today would agree with Balandin; yet few of them would be able to offer a satisfactory definition of life as an entity or a process.

In science, a hypothesis is really no more than a "let's suppose." The first Gaia book was hypothetical, and lightly written—a rough pencil sketch that tried to catch a view of the Earth seen from a different perspective. Thoughtful criticisms of this first book led to new and deeper insights into Gaia. In a physiological sense the Earth was alive. Much new evidence has accumulated, and I have made new theoretical models. We can now fill in some of the finer details, though fortunately there seems little need to erase the original lines. As a consequence this second book is a statement of Gaia theory; the basis of a new and unified view of the Earth and life sciences. Because Gaia was seen from outside as a physiological system, I have called the science of Gaia geophysiology.

Why run the Earth and life sciences together? I would ask, why have they been torn apart by the ruthless dissection of science into separate and blinkered disciplines? Geologists have tried to persuade us that the Earth is just a ball of rock, moistened by the oceans; that nothing but a tenuous film of air excludes the hard vacuum of space; and that life is merely an accident, a quiet passenger that happens to have hitched a ride on this rock ball in its journey through space and time. Biologists have been no better. They have asserted that living organisms are so adaptable that they have been fit for any material changes that have occurred during the Earth's history. But suppose that the Earth is alive. Then the evolution of the organisms and the evolution of the rocks need no longer be regarded as separate sciences to be studied in separate buildings of the university. Instead, a single evolutionary science describes the history of the whole planet. The evolution of the species and the evolution of their environment are tightly coupled together as a single and inseparable process.

Science is not obsessively concerned with whether facts are right or wrong. The practice of science is that of testing guesses; forever iterating around and towards the unattainable absolute of truth. To scientists, Gaia is a new guess that is up for trial or a novel "bioscope" through which to look at life on Earth. In some sciences, Gaian ideas are appropriate, even if not welcomed, because the vision of the world through older theories is no longer sharp and clear. This is particularly true of theoretical ecology, evolutionary biology, and the Earth sciences generally.

Theoretical ecologists for forty years—since Alfred Lotka and Vito Volterra made their simple models of a world populated only by rabbits and foxes—have tried to understand the complex interactions between a real forest and its vast range of species. Their mathematical models, though good at simulating pathologies, fail to explain the long-term stability of the complex ecosystems of the humid tropical forests. Their models seem counterintuitive; they suggest that the fragility of ecosystems increases with their diversity. They imply that the farmer who rotates his crops and keeps his hedgerows and woodland intact is not only less efficient but less ecologically stable than the monoculture factory farm.

In recent times, the evolutionary biologists have engaged in a fiery argument. The normally placid pages of those cool scientific journals, *Nature* and *Science*, have burnt like an inner city, the conservative defenders of ordered gradual change reacting against a revolution for the right to interpret Darwin's great insight. Was evolution gradual or did it proceed, as Stephen Jay Gould and Niles Eldredge propose, with long periods of stasis punctuated by catastrophic change?

Geologists interested in the evolution of the rocks, ocean, and atmosphere are beginning to ponder about the persistence of the oceans on Earth when Mars and Venus are so dry. Then there is the puzzling constancy of the climate, in spite of an ever-increasing output of heat from the Sun.

These and other things that seem obscure within their separated fields of science become clear when seen as phenomena of a living planet. Gaia theory predicts that the climate and chemical composition of the Earth are kept in homeostasis for long periods until some internal contradiction or external force causes a jump to a new stable state. On such a living planet, we shall see that punctuated evolution and abundant oceans are normal and expected.

As a theory of a living Earth, this book is neither holistic nor reductionist. There are no sections on climatology, geochemistry, and so on. The next two chapters are a statement of Gaia theory. Then follow three chapters which give a geophysiologist's view of the history of the Earth from the start of life to the present day. These run chronologically, instead of chaotically by scientific discipline. The sequence starts with the beginning of life, the Archean, when the only organisms on Earth were bacteria, and when the atmosphere was dominated by methane and oxygen was only a trace gas. Next that middle age, which the geologists call the Proterozoic, from the first appearance of oxygen as a dominant atmospheric gas until the time when communities of cells gathered to form new collectives, each with its own identity. Then a chapter on the Phanerozoic, the time of the plants and animals. In each, the record of the rocks is interpreted through Gaia theory and the new interpretation compared with the conventional wisdom of the Earth and life sciences. The final chapters concern the present and future of Gaia, with an emphasis on the human presence both on Earth and as it may one day exist on Mars. What would it take to bring Mars to life?

The arbitrariness of even a chronological division is underlined by the persistence of the Archean biota; their world has never ended, but lives on in our guts. Those bacteria have been with Gaia for nearly four thousand million years, and they still live all over the Earth in muds, sediments, and intestines—wherever they can keep away from that deadly poison, oxygen.

Any new theory about the Earth cannot be kept a secret of science. It is bound to attract the attention of humanists, environ-mentalists, and those of religious beliefs and faiths. Gaia theory is as out of tune with the broader humanist world as it is with established science. In Gaia we are just another species, neither the owners nor the stewards of this planet. Our future depends much more upon a right relationship with Gaia than with the never-ending drama of human interest.

When our family lived in the village of Bowerchalke in Wilt-shire, Helen and I would spend spring mornings seeking rare species of wild orchids. In those days, before its destruction by the machines of agribusiness vandals, the English countryside was a heavenly garden. Orchids grew in profusion on the downs, but the rarer kinds could be exceedingly hard to find. Much prior programming of the mind was needed to spot a musk orchid in the grass. It was an esoteric pastime. Much of science is done like this, and it can be enjoyable to discover new com-pounds or mathematical concepts or old ones in strange places. But these discoveries usually require rigorous mental and physical preparation and often the learning of a new language.

Gaia theory goes back to fundamentals, to genesis. Even geo-physiology is too young a science to have a language. Therefore this second book is written like the first, so that anyone interested in the idea that the Earth is alive can read it. Neither a scientific text nor the workshop manual for a planetary engineer, it is one man's view of the world where we belong. Most of all the book is for entertainment, yours and mine. It was written as part of a way of life that included time to go for walks in the country and to talk with friends, as Korolenko did, about the Earth being alive.

2

What Is Gaia?

*You must not . . . be too precise or scientific about birds and
trees and flowers . . .*

WALT WHITMAN, *Specimen Days*

Travel back in your memory to the time when you first awoke,
that exquisite moment of childhood when first you came alive—
the sudden rush of sound and sight, as if a television receiver
had been switched on and was about to bring news of vast
importance. I seem to recall sunlight and soft fresh air; then
suddenly knowing who I was and how good it was to be alive.

To reminisce about the first memory of my personal life may
seem irrelevant in our quest to understand Gaia. But it isn't.
As a scientist I observe, measure, analyze, and describe phenomena. Before I can do these things I need to know what I am
observing. In a broad sense it may be unnecessary to recognize
a phenomena when observing it, but scientists almost always
have preconceived notions of the object of their study. As a
child I recognized life intuitively. As an adult wondering about
the Earth's strange atmosphere—a mixture made of incompatible
gases such as oxygen and methane coexisting like foxes and

rabbits in the same burrow—I was forced to recognize Gaia, to intuit her existence, long before I could describe her in proper scientific terms.

The concept of Gaia is entirely linked with the concept of life. To understand what Gaia is, therefore, I first need to explore that difficult concept, life. They hate to admit it, but the life scientists, whether the natural historians of the nineteenth century or the biologists of the twentieth, cannot explain what life is in scientific terms. They all know what it is, as we have done since childhood; but in my view no one has yet succeeded in defining life. The idea of life, the sense of being alive, are the most familiar and the most difficult to understand of the concepts we meet. I have long thought that the answer to the question "What is life?" was deemed so important to our survival that it was classified "top secret" and kept locked up as an instinct in the automatic levels of the mind. During evolution, there was great selection pressure for immediate action: crucial to our survival is the instant distinction of predator from prey and kin from foe, and the recognition of a potential mate. We cannot afford the delay of conscious thought or debate in the committees of the mind. We must compute the imperatives of recognition at the fastest speed and, therefore, in the earliest-evolved and unconscious recesses of the mind. This is why we all know intuitively what life is. It is edible, lovable, or lethal.

Life as an object of scientific inquiry requiring precise definition is much more difficult. Even scientists, who are notorious for their indecent curiosity, shy away from defining life. All branches of formal biological science seem to avoid the question. In the *Dictionary of Biology* compiled by M. Abercrombie, C. J. Hickman, and M. L. Johnson, these three distinguished biologists succinctly define all manner of words like *ontogeny* (development), *Pteridophyta* (ferns), and *ecdysis* (a stage in insect development). Under the letter L there is *leptotene* (the first sign of chromosome pairing in meiosis) and *limnology* (the study of lakes), but nowhere is life mentioned. When the word *life* does appear in biology it is in rejection, as by the philosophically inclined N. W. Pirie who, in 1937, published an article entitled

"The Meaninglessness of the Terms 'Life' and 'Living'."

The Webster and the Oxford dictionaries are not much more help. Both remind of the word's origin from the Anglo-Saxon *lif*. This may explain some of the reluctance of academic biologists to tangle with so elemental a concept as life. The tribal war between the Normans and the Saxons was long enduring; the medieval schoolmen, knowing where power and preference lay, chose to support the victorious Norman establishment and to keep Latin as their language. Life was another of those rude uncivilized Anglo-Saxon words, best avoided in polite company. The Latin equivalent of lif, *anima*, was even less help. It was close in meaning to that other four-letter Gothic word, *soul*.

To go back to the Webster dictionary, it defines life as:

That property of plants and animals (ending at death and distinguishing them from inorganic matter) which makes it possible for them to take in food, get energy from it, grow, etc.

The Oxford dictionary says much the same:

The property which differentiates a living animal or plant, or a living portion of organic tissue, from dead or nonliving matter; the assemblage of the functional activities by which this property is manifested.

If such manifestly inadequate definitions of life are all I have to work with, can I do much better defining the living organism of Gaia? I have found it very difficult, but if I am to tell you about it I must try. I can start with some simpler definitions and classifications. Living things such as trees and horses and even bacteria can easily be perceived and recognized because they are bounded by walls, membranes, skin, or waxy coverings. Using energy directly from the Sun and indirectly from food, living systems incessantly act to maintain their identity, their integrity. Even as they grow and change, grow and reproduce, we do not lose track of them as visible, recognizable entities.

Although there are uncountable millions of individual organisms all growing and changing, their traits in common allow us to group them and recognize that they belong to species such as peacocks, dogs, or wheat. About ten million species are estimated to exist. When any individual fails to get energy and food, fails to act to maintain its identity, we realize it is moribund or dead.

An important step in our understanding is to recognize the significance of collections of living things. You and I are both composed of a collection of organs and tissues. The many benefi-ciaries of heart, liver, and kidney transplants testify eloquently that each of these organs can exist independently of the body when kept warm and supplied with nutrients. The organs them-selves are made up of billions of living cells, each of which can also live independently. Then the cells themselves, as Lynn Margulis has shown, are communities of microorganisms that once lived free. The energy-transforming entities of animal cells (the mitochondria) and of plants (the mitochondria and the chloroplasts) both were once bacteria living independently.

Life is social. It exists in communities and collectives. There is a useful word in physics to describe the properties of collections: *colligative*. It is needed because there is no way to express or measure the temperature or the pressure of a single molecule. Temperature and pressure, say the physicists, are the colligative properties of a sensible collection of molecules. All collections of living things show properties unexpected from a knowledge of a single one of them. We, and some other animals, keep a constant temperature whatever the temperature of our surround-ings. This fact could never have been deduced from the observa-tions of a single cell from a human being. The tendency to constancy was first noted by the French physiologist Claude Bernard in the nineteenth century. His American successor in this century, Walter Cannon, called it *homeostasis* or the wisdom of the body. Homeostasis is a colligative property of life.

We have no trouble with the idea that noble entities such as people are made up from an intricate interconnected set of cell communities. We don't find it too difficult to consider a nation or a tribe as an entity made up of its people and the territory they occupy. But what of large entities, like ecosystems and Gaia? It took the view of the Earth from space, either directly through the eyes of an astronaut, or vicariously through the visual media, to give us the personal sense of a real live planet on which the living things, the air, the oceans, and the rocks all combine in one as Gaia.

The name of the living planet, Gaia, is not a synonym for the biosphere. The biosphere is defined as that part of the Earth where living things normally exist. Still less is Gaia the same as the biota, which is simply the collection of all individual living organisms. The biota and the biosphere taken together form part but not all of Gaia. Just as the shell is part of a snail, so the rocks, the air, and the oceans are part of Gaia. Gaia, as we shall see, has continuity with the past back to the origins of life, and extends into the future as long as life persists. Gaia, as a total planetary being, has properties that are not necessarily discernible by just knowing individual species or populations of organisms living together.

The Gaia hypothesis, when we introduced it in the 1970s, supposed that the atmosphere, the oceans, the climate, and the crust of the Earth are regulated at a state comfortable for life because of the behavior of living organisms. Specifically, the Gaia hypothesis said that the temperature, oxidation state, acidity, and certain aspects of the rocks and waters are at any time kept constant, and that this homeostasis is maintained by active feedback processes operated automatically and unconsciously by the biota. Solar energy sustains comfortable conditions for life. The conditions are only constant in the short term and evolve in synchrony with the changing needs of the biota as it evolves. Life and its environment are so closely coupled that evolution concerns Gaia, not the organisms or the environment taken separately.

Most of my working life has been spent on the fringes of the life sciences, but I do not think of myself a biologist, nor do I believe would biologists accept me as one of them. When seen from outside, much of biology appears to be the building of data bases—making the "whole life catalog." Sometimes, in a pensive mood, I fancy that to biologists the living world is a vast set of book collections held in interconnected libraries. In this dream, the biologists are like competent librarians who devise the most intricate classification of every new library they discover but never read the books. They sense that something is missing from their lives, and this feeling intensifies as new collections of books grow hard to find. I see the biologists expressing an almost palpable sense of relief when joined by molecular biologists who dare to start the even greater task of classifying the words the books contain. It means that the search for the answer to the awesome question of what the books are about can be put off until the new and infinitely detailed molecular classification is complete.

My imaginary world, populated by biologists as book collectors, is in no way intended as a slur on the life sciences. Left to my own devices in such a world I should have been much less constructive. Impatient of waiting for an answer to the question, "What is the meaning of the books?" I would have seized some of them for experimental tests—for example, burning them in a calorimeter and measuring, accurately, the heat released. My sense of frustration would not have lessened when I discovered that the densely packed pages of an encyclopedia give no more heat than the same mass of plain paper. Like the biologists' classification, this physical experiment would have been profoundly unsatisfying because it would have put to Nature the wrong question.

Can we scientists, any of us, do better in our quest to understand life? There are three equally powerful approaches: molecular biology, the understanding of those information-processing chemicals that are the genetic basis of all life on Earth; physiology, the science concerned with living systems seen holistically; thermodynamics, the branch of physics that deals with time and energy and that connects living processes to the fundamental laws of the Universe. Of these sciences, the latter is the one that may go furthest in the quest to define life, yet so far has made the least progress. Thermodynamics grew from down-to-earth origins, the quest of engineers to make steam engines more efficient. It flourished in the last century, both taxing and entertaining the minds of the greatest scientists.

The first law of thermodynamics is about energy, or in other words, the capacity to do work. Energy, says the first law, is conserved. Energy in the form of sunlight falling on the leaves of a tree is used in many ways. Some is reflected so that we see the leaves as green, some is absorbed and warms them, and some is transformed to food and oxygen; ultimately, we eat the food, consume it with the oxygen we breathe, and so use the Sun's energy to move, to think, and to keep warm. The first law says that this energy is always conserved and that no matter how far it is dispersed the total always remains the same. The second law is about the dissymmetry of Nature. When heat is turned to work, some of it is wasted. The redistribution of the total quantity of energy in the Universe has direction, says the second law. It is always running down. Hot objects cool, but cool objects never spontaneously become hot. The law can appear to be broken when some metastable store of internal energy is tapped, as when a match is struck, or a piece of plutonium experiences nuclear fission, but once used up the energy cannot be recovered. The law was not broken, the energy was merely redistributed and the downward path maintained. Water does not flow up the rivers from the sea to the mountains. Natural processes always move towards an increase of disorder, and this disorder is measured by entropy. It is a quantity that always and inexorably increases.

Entropy is real, not some hazy notion invented by professors to make it easier to challenge students with difficult examination questions. Like the length of a piece of string or the temperature of wine in a glass, it is a measurable physical quantity. Indeed, like temperature, the entropy of a substance is, in a practical sense, zero at the absolute zero of $-273°C$. When heat is added to a material substance, not only the temperature increases but also the entropy. Unfortunately there is a complication: whereas temperature can be measured with a thermometer, entropy cannot be measured directly with an "entropometer." Entropy, measured in the units calories per gram per degree, is the total quantity of heat added, divided by the temperature.

Consider the lifeless perfection of a snowflake, a crystal so exquisitely ordered in its fractal pattern that it is one of the most intricate of nonliving things. The quantity of heat needed to melt a snowflake to a raindrop is 80 times larger than the quantity needed to warm the raindrop by a single degree of temperature. The increase of entropy when snowflakes melt is 80 times larger than when they warm from $-1°C$ to the melting point. Alternatively, the formation of ice that expresses the ordered perfection of a snowflake represents a decrease of entropy of the same amount. Entropy is connected in quantitative terms with the orderliness of things. The greater the order, the lower the entropy.

I like to think of entropy as the quantity that expresses the most certain property of our present Universe: its tendency to run down, to burn out. Others see it as the direction of time's arrow, a progression inevitably from birth to death. Far from being something tragic or a cause of sorrow, this universal tendency to decay benefits us. Without the decay of the Universe there could have been no Sun, and without the superabundant consumption of its energy store the Sun could never have provided the light that let us be.

The second law is the most fundamental and unchallenged law of the Universe; not surprisingly, no attempt to understand life can ignore it. The first book I read on the question of life was by the Austrian physicist, Erwin Schrödinger. He was curious about biology and wondered if the behavior of the fundamental molecules of life could be explained by physics and biology. His famous little book, entitled *What Is Life?*, is a collection of the public lectures on this topic that he gave in Dublin during his exile there in the Second World War. He describes his objective on the first page:

The large important and very much discussed question is: How can the events in space and time which take place

within the spatial boundary of a living organism be accounted
for by physics and chemistry?

He goes on to write:

The obvious inability of present-day physics and chemistry
to account for such events is no reason for doubting that
they can be accounted for by those sciences.

In those times, physicists were accustomed to exploring the
dead, near-equilibrium world of "periodic crystals"—crystals
whose regularity is predictable, one atom of one kind always
following another of a different kind in a repeating pattern.
Even these comparatively simple structures were complex
enough to stretch to the limit the simple equipment then available. Organic chemists were discovering the intricate structures
of the "aperiodic crystals" from living matter, such as the proteins, polysaccharides, and nucleic acids. They were still far
from the present-day understanding of the chemical nature of
genetic material. Schrödinger concluded that, metaphorically,
the most amazing property and characteristic of life is its ability
to move upstream against the flow of time. Life is the paradoxical
contradiction to the second law, which states that everything
is, always has been, and always will be running down to equilibrium and death. Yet life evolves to ever-greater complexity and
is characterized by an omnipresence of improbability that would
make winning a sweepstake every day for a year seem trivial
by comparison. Even more remarkable, this unstable, this apparently illegal, state of life has persisted on the Earth for a sizable
fraction of the age of the Universe itself. In no way does life
violate the second law; it has evolved with the Earth as a tightly
coupled system so as to favor survival. It is like a skilled accoun-
tant, never evading the payment of required tax but also never
missing a loophole. Most of Schrödinger's book is an optimistic
prediction of how life is knowable. The eminent molecular biologist, Max Perutz, has recently commented that little in Schrö-

dinger's book is original, and what is original is often wrong. This may be true; but I, like many of my colleagues, still acknowledge a debt to Schrödinger for having set us thinking in a productive way.

The great physicist Ludwig Boltzmann expressed the meaning of the second law in an equation of great seemliness and simplicity: $S = k(lnP)$, where S is that strange quantity entropy; k is a constant rightly called the Boltzmann constant; and lnP is the natural logarithm of the probability. It means what it says—the less probable something is, the lower is its entropy. The most improbable thing of all, life, is therefore to be associated with the lowest entropy. Schrödinger was not happy to associate something as significant as life with a diminished quantity, entropy. He proposed, instead, the term "negentropy," the reciprocal of entropy—that is, 1 divided by entropy or $1/S$. Negentropy is large, of course, for improbable things like living organisms. To describe the burgeoning life of our planet as improbable may seem odd. But imagine that some cosmic chef takes all the ingredients of the present Earth as atoms, mixes them, and lets them stand. The probability that those atoms would combine into the molecules that make up our living Earth is zero. The mixture would always react chemically to form a dead planet like Mars or Venus.

Often in science the same idea is thought of in different contexts in different parts of the world. There is nothing occult about this. Ideas are in continuous use as currency in the exchanges between scientists and, like money, can be used to buy many different things. When Schrödinger was lecturing about negentropy in Dublin, Claude Shannon was investigating a similar quantity in the United States, but from a radically different perspective. Shannon, at the Bell Telephone Laboratories, was developing information theory. It started as a plain engineering quest to discover the physical factors that caused a message sent by cable or by radio to lose information as it passed from the sender to the receiver. Shannon soon discovered a quantity that always tended to increase; the size of the increase was a measure of the loss of information. In no experiment was the size of this quantity ever observed to decrease. On the advice of John Von Neumann, a mathematical physicist, Shannon named this quantity entropy because it so much resembled the entropy of the steam engineers. The reciprocal of Shannon's entropy is the quantity often called information. If we assume that the entropy Shannon discovered is the same as the entropy of the steam engineers, then the elusive quantity that Schrödinger associated with the improbability of life—negentropy—is comparable with Shannon's information. In mathematical terms, if S is the entropy then both negentropy and information are $1/S$.

The reward that comes from persevering with thoughts about these difficult concepts is insight to illuminate our quest to understand life and Gaia. The contribution from Shannon's theory is that information is not just knowledge. Information, in thermodynamic terms, is a measure of the absence of ignorance. Better to know all about a simple system than merely a great deal about a complex one. The less the ignorance, the lower the entropy. This is why it is so difficult to grasp the concept of Gaia from the voluminous but isolated knowledge of a single scientific discipline.

If the second law tells us that entropy in the Universe is increasing, how does life avoid the universal tendency for decay? A physicist in Britain, J. D. Bernal, tried to balance the books. In 1951, he wrote in recondite terminology: "Life is one member of the class of phenomena which are open or continuous reaction systems able to decrease their internal entropy at the expense of free energy taken from the environment and subsequently rejected in degraded form." Many other scientists have expressed these words as a mathematical equation. Among the clearest and most readable are the statements in a small book, *The Thermodynamics of the Steady State,* written by a physical chemist, K. G. Denbigh. They can be restated less rigorously but more comprehensibly as follows. By the act of living, an organism continuously creates entropy and there will be an outward flux of entropy across its boundary. You, as you read these words, are creating entropy by consuming oxygen and the fats and sugars stored in your body. As you breathe, you excrete waste products high in entropy into the air, such as carbon dioxide, and your warm body emits to your surroundings infrared radiation high in entropy. If your excretion of entropy is as large or larger than your internal generation of entropy, you will continue to live and remain a miraculous, improbable, but still legal avoidance of the second law of the Universe. "Excretion of entropy" is just a fancy way of expressing the dirty words excrement and pollution. At the risk of having my membership card of the Friends of the Earth withdrawn, I say that only by pollution do we survive. We animals pollute the air with carbon dioxide, and the vegetation pollutes it with oxygen. The pollution of one is the meat of another. Gaia is more subtle, and, at least until humans appeared, polluted this region of the Solar System with no more than the gentle warmth of infrared radiation.

In recent times, some interesting insights have come from the investigations of Ilya Prigogine and his colleagues into the thermodynamics of eddies, vortices, and many other transient systems that are low in entropy. Things like eddies and whirlpools develop spontaneously when there is a sufficient flux of free energy. It was in the nineteenth century that a British physicist, Osborne Reynolds, curious about the conditions that led to turbulence in the flow of fluids, discovered that the onset of eddies in a stream or in a flow of gas takes place only when the flow exceeds a critical value. A useful analogy here is that if you blow a flute too gently no sound emerges. But if you blow hard enough, wind eddies form and are made part of the system that makes sound. Extending the earlier mathematics of the American physical chemist Lars Onsager, Prigogine and his colleagues have applied the thermodynamics of the steady state to develop what might be called the thermodynamics of the "unsteady state." They classify these phenomena by the term "dissipative structures." They have structure, but not the permanency of solids; they dissipate when the supply of energy is turned off. Living organisms include dissipative structures within them, but the class is broadly based. It includes many manufactured things, such as refrigerators, and natural phenomena such as flames, whirlpools, hurricanes, and certain peculiar chemical reactions. Living things are so infinitely complex in comparison with the dissipative structures of the fluid state that many feel that, although on the right track, present-day thermodynamics has far to go in defining life. Physicists, chemists, and biologists, although not rejecting these notions, do not make them part of the inspiration of their working lives. Their response is like that of a wealthy congregation to the exhortations of their priest on the virtues of poverty. It is something felt to be good, but not a way of life for next week.

A crucial insight that comes from Schrödinger's generalizations about life is that living systems have boundaries. Living organisms are open systems in the sense that they take and excrete energy and matter. In theory, they are open as far as the bounds of the Universe; but they are also enclosed within a hierarchy of internal boundaries. As we move in towards the Earth from space, first we see the atmospheric boundary that encloses Gaia; then the borders of an ecosystem such as the forests; then the skin or bark of living animals and plants; further in are the cell membranes; and finally the nucleus of the cell and its DNA. If life is defined as a self-organizing system characterized by an actively sustained low entropy, then, viewed from outside each of these boundaries, what lies within is alive.

You may find it hard to swallow the notion that anything as large and apparently inanimate as the Earth is alive. Surely, you may say, the Earth is almost wholly rock and nearly all incandescent with heat. I am indebted to Jerome Rothstein, a physicist, for his enlightenment on this, and other things. In a thoughtful paper on the living Earth concept (given at a symposium held in the summer of 1985 by the Audubon Society) he observed that the difficulty can be lessened if you let the image of a giant redwood tree enter your mind. The tree undoubtedly is alive, yet 99 percent is dead. The great tree is an ancient spire of dead wood, made of lignin and cellulose by the ancestors of the thin layer of living cells that go to constitute its bark. How like the Earth, and more so when we realize that many of the atoms of the rocks far down into the magma were once part of the ancestral life from which we all have come.

When the Earth was first seen from outside and compared as a whole planet with its lifeless partners, Mars and Venus, it was impossible to ignore the sense that the Earth was a strange and beautiful anomaly. Yet this unconventional planet probably would have been kept in the scullery, like Cinderella, had not NASA in the role of Prince offered a rescue by way of the planetary exploration program. As we saw in chapter 1, the questions raised by space science were at first narrowly focused on a practical question: How is life on another planet to be recognized? Because that question could not be explained solely by conventional biology or geology, I became preoccupied with another question: What if the difference in atmospheric composition between the Earth and its neighbors Mars and Venus is a consequence of the fact that the Earth alone bears life?

The least complex and most accessible part of a planet is its atmosphere. Long before the Viking spacecraft landed on Mars, or the Russian Venera landed on Venus, we knew the chemical compositions of their atmospheres. In the middle 1960s, telescopes tuned to the infrared radiation reflected by the molecules of atmospheric gases were used to view Mars and Venus. These observations revealed the identity and proportion of the gases with fair accuracy. Mars and Venus both had atmospheres dominated by carbon dioxide, with only small proportions of oxygen and nitrogen. More important, both had atmospheres close to the chemical equilibrium state; if you took a volume of air from either of those planets, heated it to incandescence in the presence of a representative sample of rocks from the surface, and then allowed it to cool slowly, there would be little or no change in composition after the experiment. The Earth, by contrast, has an atmosphere dominated by nitrogen and oxygen. A mere trace of carbon dioxide is present, far below the expectation of planetary chemistry. There are unstable gases such as nitrous oxide, and gases such as methane that react readily with the abundant oxygen. If the same heating-and-cooling experiment were tried with a sample of the air that you are now breathing, it would be changed. It would become like the atmospheres of Mars and Venus: carbon dioxide dominant, oxygen and nitrogen greatly diminished, and gases such as nitrous oxide and methane absent. It is not too far-fetched to look on the air as like the gas mixture that enters the intake of an internal combustion engine: combustible gases, hydrocarbons, and oxygen mixed. The atmospheres of Mars and Venus are like the exhaust gases, all energy spent.

The amazing improbability of the Earth's atmosphere reveals negentropy and the presence of the invisible hand of life. Take for example oxygen and methane. Both are present in our atmosphere in constant quantities; yet in sunlight they react chemically to give carbon dioxide and water vapor. Anywhere you travel on the Earth's surface to measure it, the methane concentration is about 1.5 parts per million. Close to 1,000 million tons of methane must be introduced into the atmosphere annually to maintain methane at a constant level. In addition, the oxygen used in oxidizing the methane must be replaced—at least 2,000 million tons yearly. The only feasible explanation for the persistence of this unstable atmosphere at a constant composition, and for periods vastly longer than the reaction times of its gases, is the influence of a control system, Gaia.

It is often difficult to recognize the larger entity of which we are a part; as the saying goes, "You can't see the forest for the trees." So it was with the Earth itself before we shared with the astronauts vicariously that stunning and awesome vision; that impeccable sphere that punctuates the division of the past from the present. This gift, this ability to see the Earth from afar, was so revealing that it forced the novel top-down approach to planetary biology. The conventional wisdom of biology on Earth itself had always been forced to take a bottom-up approach by the sheer size of the Earth when compared with us or any living thing we knew. The two approaches are complementary. In the understanding of a microbe, an animal, or a plant, the top-down physiological view of life as a whole system harmoniously merges with the bottom-up view originating with molecular biology: that life is an assembly made from a vast set of ultramicroscopic parts. Since James Hutton there has been a "loyal opposition" of scientists who doubted the conventional wisdom that the evolution of the environment is determined by chemical and physical forces alone. Vernadsky adopted Suess's concept of the biosphere to define the boundaries of the realm of the biota. Since Vernadsky, there has been a continuous tradition (called biogeochemistry) in the Soviet Union—and, to a lesser extent, elsewhere—that has recognized the interaction between the soils, oceans, lakes, and rivers and the life they bear. It is well stated by a Russian, M. M. Yermolaev, in *An Introduction to Physical Geography*: "The biosphere is understood as being that part of the geographical envelope of the Earth, within the boundaries of which the physico-geographical conditions ensure the normal work of the enzymes." More recent members of this scientific opposition have included the following: Alfred Lotka of John Hopkins University, and Eugene Odum, who alone among ecologists took a physiological view of ecosystems; two Americans of European origin, the limnologist G. Evelyn Hutchinson and the paleontologist Heinz A. Lowenstam; an oceanographer from Britain, A. Redfield; and a Swedish geochemist, L. G. Sillén. They all have recognized the importance of the participation by life in the evolution of the environment. Most geologists, however, have neglected the presence of living organisms as an active participant in their theories of the Earth's evolution.

The counterpart of this geological apartheid is the failure of most biologists to recognize that the evolution of the species is strongly coupled with the evolution of their environment. For example, in 1982 there appeared a book, *Evolution Now: A Century after Darwin*, edited by John Maynard Smith, which consisted of a collection of essays by distinguished biologists on the most controversial issues of evolutionary biology. In this collection, the only (and enigmatic) mention of the environment is in an essay by Stephen Jay Gould: "Organisms are not billiard balls, struck in a deterministic fashion by the cue of natural selection and rolling to optimal positions on life's table. They influence their own destiny in interesting and complex and comprehensible ways. We must put this concept of organism back into evolutionary biology."

Apart from Lynn Margulis, the only other biologist I know to have taken the environment into account when considering life is J. Z. Young. In 1971, this distinguished physiologist was independently moved to write in a chapter on homeostasis in his book, *An Introduction to the Study of Man:* "The entity that is maintained intact, and of which we all form a part, is not the life of one of us, but in the end the whole of life upon the planet." J. Z. Young's view serves as a link between Gaia theory and the general scientific consensus. Through Gaia theory, I see the Earth and the life it bears as a system, a system that has the capacity to regulate the temperature and the composition of the Earth's surface and to keep it comfortable for living organisms. The self-regulation of the system is an active process driven by the free energy available from sunlight.

The early reaction, soon after the Gaia hypothesis was introduced in the early 1970s, was ignorance in the literal sense. For the most part the Gaian idea was ignored by professional scientists. It was not until the late 1970s that it was subjected to criticism.

Good criticism is like bathing in an ice-cold sea. The sudden chill of immersion in what seems at first a hostile medium soon stirs the blood and sharpens the senses. My first reaction on reading W. Ford Doolittle's criticism of the Gaia hypothesis in *CoEvolution Quarterly* in 1979 was shock and incoherent disbelief. The article was splendidly put together and beautifully written, but this did not lessen its frigidity. Icy waters may be pellucid, but this does not make them warm. After an icy plunge, however, comes that warm sense of relaxation when sunning on the beach. After a while, I began to realize that Ford Doolittle's criticism could be taken not so much as an attack on Gaia but as a criticism of the inadequacy of its presentation.

Gaia had first been seen from space and the arguments used were from thermodynamics. To me it was obvious that the Earth was alive in the sense that it was a self-organizing and self-regulating system. To Ford Doolittle, from his world of molecular biology, it was equally obvious that evolution by natural selection could never lead to "altruism" on a global scale. He was supported in the similarly forceful and effective writings of Richard Dawkins in his book, *The Extended Phenotype* (1982). From their world of microscopes, how could the "selfish" interests of living cells be expressed at the distance of a planet? For these competent and dedicated biologists, positing the regulation of the atmosphere by microbial life seemed as absurd as expecting the legislation of some human government to affect the orbit of Jupiter. I am indebted to them both for having shown clearly that we were taking far too much for granted, and that Gaia lacked a firm theoretical basis.

Not only did molecular biologists object to Gaia. Two other valued critics were the climatologist Stephen Schneider from Colorado, and the geochemist H. D. Holland from Harvard. They, in common with most of their peers, preferred to explain the facts of the evolution of the rocks, the ocean, the air, and the climate by chemical and physical forces alone. In his book *The Chemical Evolution of the Atmosphere and the Oceans,* Holland wrote: "I find the hypothesis intriguing and charming, but ultimately unsatisfactory. The geologic record seems much more in accord with the view that the organisms that are better able to compete have come to dominate, and that the Earth's near surface environment and processes have accommodated themselves to changes wrought by biological evolution. Many of these changes must have been fatal or near fatal to parts of the contemporary biota. We live on an Earth that is the best of all worlds but only for those who have adapted to it." Stephen Schneider's objection—expressed in his book with Randi Londer, *The Coevolution of Climate and Life*—was to the implication in the early papers on Gaia that homeostasis was the only means of climate regulation. I am indebted to all of these critics for having shown clearly that we were taking too much for granted, and that Gaia lacked a firm theoretical basis. Greater than this is my gratitude to Stephen Schneider who made sure that Gaia was properly debated by the scientific community by calling a Chapman Conference of the American Geophysical Union in March 1988.

To many scientists Gaia was a teleological concept, one that required foresight and planning by the biota. How in the world could the bacteria, the trees, and the animals have a conference to decide optimum conditions? How could organisms keep oxygen at 21 percent and the mean temperature at 20°C? Not seeing a mechanism for planetary control, they denied its existence as a phenomenon and branded the Gaia hypothesis as teleological. This was a final condemnation. Teleological explanations, in academe, are a sin against the holy spirit of scientific rationality; they deny the objectivity of Nature.

But when making this severest of criticisms of Gaia, the scientists may not have noticed the extent of their own errors. The innocent use of that slippery concept "adaptation" is another path to damnation. Earth is indeed the best of all worlds for those who are adapted to it. But the excellence of our planet takes on a different significance in the light of the evidence that geochemists themselves have gathered. Evidence that shows the Earth's crust, oceans, and air to be either directly the product of living things or else massively modified by their presence. Consider how the oxygen and nitrogen of the air come directly from plants and microorganisms, and how the chalk and limestone rocks are the shells of living things once floating in the sea. Life has not adapted to an inert world determined by the dead hand of chemistry and physics. We live in a world that has been built by our ancestors, ancient and modern, and which is continuously maintained by all things alive today. Organisms are adapting in a world whose material state is determined by the activities of their neighbors; this means that changing the environment is part of the game. To think otherwise would require that evolution was a game with rules like cricket or baseball—one in which the rules forbad environmental change. If, in the real world, the activity of an organism changes its material environment to a more favorable state, and as a consequence it leaves more progeny, then both the species and the change will increase until a new stable state is reached. On a local scale adaptation is a means by which organisms can come to terms with unfavorable environments, but on a planetary scale the coupling between life and its environment is so tight that the tautologous notion of "adaptation" is squeezed from existence. The evolution of the rocks and the air and the evolution of the biota are not to be separated.

It is a tribute to the success of biogeochemistry that most Earth scientists today agree that the reactive gases of the atmosphere are biological products. But most would disagree that the biota in any way control the composition of the atmosphere, or any of the important variables, such as global temperature and oxygen concentration, which depend on the atmosphere. There are two principal objections to Gaia, first that it is teleological, and that for the regulation of the climate, the chemical composition on a planetary scale, a kind of forecasting, a clairvoyance, would be needed. The second objection, most clearly expressed by Stephen Schneider, is that biological regulation is only partial, and that the real world is a "coevolution" of life and the inorganic. The second criticism is the more difficult, and in many ways the purpose of this book is to try to answer it. The first, the teleological criticism, I think is wrong and I will now try to show why.

I knew that there was little point in gathering more evidence about the now-obvious capacity of the Earth to regulate its climate and composition. Mere evidence by itself could not be expected to convince mainstream scientists that the Earth was regulated by life. Scientists usually want to know how it works; they want a mechanism. What was needed was a Gaian model. In those hybrid sciences of biogeochemistry and biogeophysics, models of environmental change do not permit a regulatory role to the biota. The practitioners of these sciences assume that the operating points of the system are fixed by chemical and physical properties. For example, snow melts or forms at 0°C. The reflection of sunlight by snow cover can provide a powerful positive feedback on cooling, and a system for regulating the climate could be based on the melting or formation of snow. But there is no way for the melting point of snow, which is a characteristic of ice as a substance, to change to a more comfortable warmth of, say, 20°C. In great contrast, the operating points of a living organism are always set at favorable levels.

In what way do Gaian models differ from the conventional biogeochemical ones? Does the assumption of the close coupling of life and its environment change the nature of the whole system? Is homeostasis a reasonable prediction of Gaia theory? The difficulty in answering these questions comes from the sheer complexity of the biota and the environment, and because they are interconnected in multiple ways. There is hardly a single aspect of their interaction that we can confidently describe by a mathematical equation. A drastic simplification was needed. I wrestled with the problem of reducing the complexity of life and its environment to a simple scheme that could enlighten without distorting. Daisyworld was the answer. I first described this model in 1982 at a conference on biomineralization in Amsterdam, and published a paper, "The Parable of Daisyworld," in *Tellus* in 1983 with my colleague Andrew Watson. I am indebted to Andrew for the clear, graphic way of expressing it in formal mathematical terms in this paper.

Picture a planet about the same size as the Earth, spinning on its axis and orbiting, at the same distance as the Earth, a star of the same mass and luminosity as the Sun. This planet differs from the Earth in having more land area and less ocean, but it is well watered, and plants will grow almost anywhere on the land surfaces when the climate is right. This is the planet Daisyworld, so called because the principal plant species are daisies of different shades of color: some dark, some light, and some neutral colors in between. The star that warms and illuminates Daisyworld shares with our Sun the property of increasing its output of heat as it ages. When life started on Earth some 3.8 billion years ago, the Sun was about 30 percent less luminous than now. In a few more billion years, it will become so fiercely hot that all life that we know will die or be obliged to find another home planet. The increase of the Sun's brightness as it ages is a general and undoubted property of stars. As the star burns hydrogen (its nuclear fuel) helium accumulates. The helium, in the form of a gaseous ash, is more opaque to radiant energy than is hydrogen and so impedes the flow of heat from the nuclear furnace at the center of the star. The central tempera-ture then rises and this in turn increases the rate of hydrogen burning until there is a new balance between heat produced at the center and the heat lost from the solar surface. Unlike ordinary fires, star-sized nuclear fires burn fiercer as the ash accumulates and sometimes even explode.

Daisyworld is simplified, reduced if you like, in the following ways. The environment is reduced to a single variable, tempera-ture, and the biota to a single species, daisies. If too cold, below 5°C, daisies will not grow; they do best at a temperature near 20°C. If the temperature exceeds 40°C, it will be too hot for the daisies, and they will wilt and die. The mean temperature of the planet is a simple balance between the heat received from the star and the heat lost to the cold depths of space in the form of long-wave infrared radiation. On the Earth, this heat balance is complicated by the effects of clouds and of gases such as carbon dioxide. The sunlight may be reflected back to space by the clouds before it can reach and warm the surface. On the other hand, the heat loss from the warm surface may be lessened because clouds and molecules of carbon dioxide reflect it back to the surface. Daisyworld is assumed to have a constant amount of carbon dioxide, enough for daisies to grow but not so much as to complicate the climate. Similarly, there are no clouds in the daytime to mar the simplicity of the model, and all rain falls during the night.

The mean temperature of Daisyworld is, therefore, simply determined by the average shade of color of the planet, or as astronomers call it, the albedo. If the planet is a dark shade, low albedo, it absorbs more heat from the sunlight and the surface is warmed. If light in color, like fallen snow, then 70 or 80 percent of the sunlight may be reflected back to space. Such a surface is cold when compared with a dark surface under comparable solar illumination. Albedos range from 0 (wholly black) to 1 (wholly white). The bare ground of Daisyworld is usually taken to have an albedo of 0.4 so that it absorbs 40 percent of the sunlight that falls upon it. Daisies range in shade of color from dark (with an albedo of 0.2) to light (with an albedo of 0.7).

Imagine a time in the distant past of Daisyworld. The star that warms it was less luminous, so that only in the equatorial region was the mean temperature of bare ground warm enough, 5°C, for growth. Here daisy seeds would slowly germinate and flower. Let us assume that in the first crop multicolored, light, and dark species were equally represented. Even before the first season's growth was over, the dark daisies would have been favored. Their greater absorption of sunlight in the localities where they grew would have warmed them above 5°C. The light-colored daisies would be at a disadvantage. Their white flowers would have faded and died because, reflecting the sunlight as they do, they would have cooled below the critical temperature of 5°C.

The next season would see the dark daisies off to a head start, for their seeds would be the most abundant. Soon their presence would warm not just the plants themselves, but, as they grew and spread across the bare ground, would increase the temperature of the soil and air, at first locally and then regionally. With this rise of temperature, the rate of growth, the length of the warm season, and the spread of dark daisies would all exert a positive feedback and lead to the colonization of most of the planet by dark daisies. The spread of dark daisies would eventually be limited by a rise of global temperature to levels above the optimum for growth. Any further spread of dark daisies would lead to a decline in seed production. In addition, when the global temperature is high, white daisies will grow and spread in competition with the dark ones. The growth and spread of white daisies is favored then because of their natural ability to keep cool.

As the star that shines on Daisyworld grows older and hotter, the proportion of dark to light daisies changes until, finally, the heat flux is so great that even the whitest daisy crop cannot keep enough of the planet below the critical 40°C upper limit for growth. At this time flower power is not enough. The planet

becomes barren again, and so hot that there is no way for daisy
life to start again.

It is easy to make a numerical model of Daisyworld simple
enough to run on a personal computer. Daisy populations are
modeled by differential equations borrowed from theoretical ecology (Carter and Prince, 1981). The mean temperature of the
planet is calculated directly from the balance of the heat it
receives from its star and the heat it loses by radiation to the
cold depths of space. Figure 2.1 shows the evolution of the

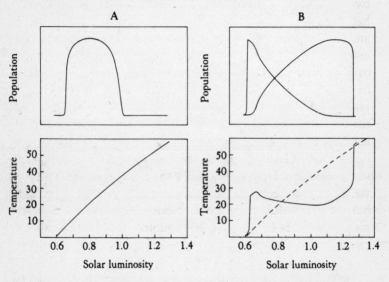

2.1 Models of the evolution of Daisyworld according to conventional
wisdom (A) and to geophysiology (B). The upper panels illustrate daisy
populations in arbitrary units; the lower panels, temperatures in degrees
Celsius. Going from left to right along the horizontal axis, the star's
luminosity increases from 60 to 140 percent of that of our own Sun.
A illustrates how the physicists and the biologists in complete isolation
calculate their view of the evolution of the planet. According to this
conventional wisdom, the daisies can only respond or adapt to changes
in temperature. When it becomes too hot for comfort, they will die. But
in the Gaian Daisyworld (B), the ecosystem can respond by the competitive
growth of the dark and light daisies, and regulate the temperature over a
wide range of solar luminosity. The dashed line in the lower panel in B
shows how the temperature would rise on a lifeless Daisyworld.

temperature and the growth of daisies during the progressive increase in heat flux from its star according to the conventional wisdom of physics and biology, and according to geophysiology.

When I first tried the Daisyworld model I was surprised and delighted by the strong regulation of planetary temperature that came from the simple competitive growth of plants with dark and light shades. I did not invent these models because I thought that daisies, or any other dark- and light-colored plants, regulate the Earth's temperature by changing the balance between the heat received from the Sun and that lost to space. I had designed them to answer the criticism of Ford Doolittle and Richard Dawkins that Gaia was teleological. In Daisyworld, one property of the global environment, temperature, was shown to be regulated effectively, over a wide range of solar luminosity, by an imaginary planetary biota without invoking foresight or planning. This is a definitive rebuttal of the accusation that the Gaia hypothesis is teleological, and so far it remains unchallenged.

So what is Gaia? If the real world we inhabit is self-regulating in the manner of Daisyworld, and if the climate and environment we enjoy and freely exploit is a consequence of an automatic, but not purposeful, goal-seeking system, then Gaia is the largest manifestation of life. The tightly coupled system of life and its environment, Gaia, includes:

1. Living organisms that grow vigorously, exploiting any environmental opportunities that open.

2. Organisms that are subject to the rules of Darwinian natural selection: the species of organisms that leave the most progeny survive.

3. Organisms that affect their physical and chemical environment. Thus animals change the atmosphere by breathing: taking in oxygen and letting out carbon dioxide. Plants and algae do the reverse. In numerous other ways, all forms of life incessantly modify the physical and chemical environment.

4. The existence of constraints or bounds that establish the limits of life. It can be too hot or too cold; there is a comfortable warmth in between, the preferred state. It can be too acid or too alkaline; neutrality is preferred. Almost all chemicals have a range of concentrations tolerated or needed by life. For many elements, such as iodine, selenium, and iron, too much is a poison, too little causes starvation. Pure uncontaminated water will support little; but neither will the saturated brine of the Dead Sea.

Few scientists would object to any of these four conditions, either singly or taken as a group. When they are taken together as a tightly coupled ensemble, they seem to form a recipe for a Gaian system. The ensemble is a fruitful source of models of self-regulating systems like Daisyworld. The fourth condition, which sets the physical and chemical bounds of life, I find the most interesting, unexpected, and full of insight. One has only to think of the social analogue of the family or community that exists with firm but reasonable bounds in comparison with one in which the limits of behavior are ill-defined. Stability and well-defined bounds seem to go together. Physicists are agreed that life is an open system. But like one of those Russian dolls which enclose a series of smaller and still smaller dolls, life exists within a set of boundaries. The outer boundary is the Earth's atmospheric edge to space. Within the planetary boundary, entities diminish but grow ever more intense as the inward progression goes from Gaia to ecosystems, to plants and animals, to cells and to DNA. The boundary of the planet then circumscribes a living organism, Gaia, a system made up of all the living things and their environment. There is no clear distinction anywhere on the Earth's surface between living and nonliving matter. There is merely a hierarchy of intensity going from the "material" environment of the rocks and the atmosphere to the living cells. But at great depths below the surface, the effects of life's presence fade. It may be that the core of our planet is unchanged as a result of life; but it would be unwise to assume it.

In exploring the question, "What is life?" we have made some progress. By looking at life through Gaia's telescope, we see it as a planetary-scale phenomenon with a cosmological life span. Gaia as the largest manifestation of life differs from other living organisms of Earth in the way that you or I differ from our population of living cells. At some time early in the Earth's history before life existed, the solid Earth, the atmosphere, and oceans were still evolving by the laws of physics and chemistry alone. It was careering, downhill, to the lifeless steady state of a planet almost at equilibrium. Briefly, in its headlong flight through the ranges of chemical and physical states, it entered a stage favorable for life. At some special time in that stage, the newly formed living cells grew until their presence so affected the Earth's environment as to halt the headlong dive towards equilibrium. At that instant the living things, the rocks, the air, and the oceans merged to form the new entity, Gaia. Just as when the sperm merges with the egg, new life was conceived.

The quest to define life might be compared with assembling a jigsaw puzzle, a puzzle where a landscape scene is cut into a thousand small interlocking pieces and the pieces scrambled. Classification is needed to put it together again. The blue sky is easy to separate from the brown earth and green trees. Skilled solvers of the jigsaw puzzle know that a key step is to find and connect the straight-sided pieces that define the edge, the boundary of the scene. The discovery that the outer reaches of the atmosphere are a part of planetary life in a like manner has defined the edge of our puzzle picture of the Earth. Once the edge is completely assembled, at least the size of the picture is known and the placing of the inner groupings made easier. Gaia is no static picture. She is forever changing as life and the Earth evolve together, but in our brief life span she keeps still long enough for us to begin to understand and see how fair she is.

3

Exploring
Daisyworld

*Give me a fruitful error any time, full of seeds, bursting with
its own corrections.*
VILFREDO PARETO, *comment on Kepler*

The word *theory* has the same Greek root as *theatre*. Both are
concerned with putting on a show. A theory in science is no
more than what seems to its author a plausible way of dressing
up the facts and presenting them to the audience. Like plays,
theories are judged according to several different, and barely
connected, criteria. Artistic content is important; a theory that
is elegant, inspiring, and presented with craftsmanship is univer-
sally appreciated. But hard-working scientists like best theories
that are full of predictions which can easily be tested. It matters
little whether the view of the theorizer is right or wrong: investigation and research are stimulated, new facts discovered, and
new theories composed. That it was wrong did little to detract
from the theory of continuous creation put forward by the astronomers Hoyle, Bondi, and Gold. It has now been abandoned,
but in its day it was a deeply satisfying intellectual concept.
The only bad theories are those that cannot be questioned or

tested. What use is a theory that the Universe was created, complete with inhabitants, and all with memories of a non-existent past, at 15.37 hr GMT on October 27, 1917? There is no way to prove or disprove it, and it makes no useful predictions.

At first glance, Gaia theory might seem to be untestable. Obviously, it would be improper and irresponsible to attempt vivisection on a whole living planet. The nineteenth-century "blood up to the elbows" school of investigating living things is passé. We have learnt from engineers, who value their contraptions more than most of us value the infinitely more complex and beautiful mechanisms of living organisms, that so much can be learnt from the noninvasive testing of a system that vivisection is not needed. In many different ways, Gaia theory is wide open for experimental investigation.

The most direct evidence comes from the real world as it now is. Just as we can observe the pulse, the blood pressure, the electrical activity of the heart, and so on without interfering with the normal physiology of a human subject; so we can observe the circulation of air, the oceans, and the rocks. We can measure the seasonal pulsing of the carbon dioxide of the air as the plants breathe it in and the consumers breathe it out. We can follow the course of essential nutrients from the rocks to the ocean to the air and back again, and see how at each step different but interlinked systems are affected.

There is also a vast amount of historical evidence preserved in the rocks. During its life span, our planet has suffered the impact of planetesimals. We have been hit by close to thirty small planets, up to 10 miles in diameter and traveling as fast as sixty times the speed of sound. These impacts release about a thousand times as much energy as would be released if all the nuclear powers exploded all the present weapon stocks. Such events do more than make 200-mile craters, they can destroy up to 90 percent of all living organisms from the microscopic to the macroscopic. The impacts make the Earth ring like a bell, and the reverberations of the event resonate, metaphorically, throughout the systems of the Earth for maybe a million years

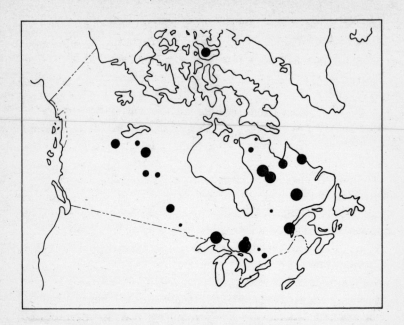

3.1 Map of Canada showing large meteor craters.

or more. The history of our planet is punctuated by these perturbations; from their record we can learn a great deal about how
the system works and the way that homeostasis is fully restored.
Should you doubt that the Earth has been so stricken so often,
glance at the map of the distribution of craters on the older
surface rocks in Canada (figure 3.1). It is like a glance at a
region of the Moon's surface. On most of the continental areas
and on all the surface of the sea floor, however, the rapidity
of the smoothing processes of weathering and sea-floor spreading
rapidly remove the evidence of any but the most recent impacts.

Not all catastrophic events are from external causes; some,
such as the appearance of oxygen gas, are generated by inherent
internal contradictions within the system and can be likened
to such crises in living organisms as puberty, menopause, or
the metamorphosis of a pupa to a butterfly. The record of the
rocks, though blurred by time and often incomplete, still pre-
serves some evidence of the chemical and physical state of the

Earth and of the distribution of the species before and after each of these perturbations. But disentangling the record is rather like trying to find traces of the identity of a terrorist from the rubble of the building his bomb destroyed.

The most persuasive criticism of Gaia theory is that planetary homeostasis, by and for living organisms, is impossible because it would require the evolution of communication between the species and a capacity for foresight and planning. The critics who made this challenging, and to me helpful, criticism were not concerned with the practical evidence that the Earth has kept a climate favorable for life in spite of major perturbations, or that the atmosphere is now stable in its composition in spite of the chemical incompatibility of its component gases. They were criticizing from the certainty of their knowledge of biology. No organism as large, and, as they saw her, sentient, could possibly exist. I think this criticism is dogmatic, and, as we saw in the last chapter, it is easy to answer. The simple model Daisyworld illustrated how Gaia might work. It pictured an imaginary world that spun like the Earth as it circled and was warmed by a star that was the identical twin of our own Sun. On this world, the competition for territory between two species of daisies, one dark and one light in color, led to the accurate regulation of planetary temperature close to that comfortable for plants like daisies. No foresight, planning, or purpose was invoked. Daisyworld is a theoretical view of a planet in homeostasis. We can now begin to think of Gaia as a theory, something rather more than the mere "let's suppose" of an hypothesis.

There is much more to Daisyworld than just the answer to a criticism. I first made it for that purpose, but as it has developed I have found it to be a source of insight and an answer to questions about theoretical ecology and Darwinism, as well as to questions about Gaia. An important property of the model is its docility and stability in mathematical terms. As I continue to work with these models I find that the number of species that can be accommodated appears to be limited only by the speed and capacity of the computer used and by my patience. Whatever the details, the inclusion of feedback from the environment appears to stabilize the system of differential equations used to model the growth and competition of the species. Most of what follows is the record of my explorations in Daisyworld and an account of the discoveries there. I have assumed that many of my readers are not moved by mathematical expressions, and have therefore not included these. For those who regard any theory not expressed in the pure language of mathematics as at best inadequate, Andrew Watson and I have described the mathematics of Daisyworld in our paper in *Tellus*. In no way is the stability of Daisyworld dependent upon an idiosyncratic choice of initial values, or rate constants, and as we shall see in later chapters the model is general in its application.

It may still be that some of the diagrams used to illustrate the geophysiology of the model planet are difficult to follow for readers unfamiliar with this form of graphic explanation. For you, I have written this book so that this chapter can be skipped without much loss, provided that you are already convinced that Gaia theory gives a fair representation of the Earth. But I ask my critics to read on, for here I shall try to answer in detail the objections they have raised.

The reaction of scientists to the Daisyworld model was revealing. Meteorologists and climatologists were the most interested; geologists and geochemists next. With rare exceptions, biologists either ignored the models or remained as skeptical as ever. A persistent criticism from biologists was that, in a real world, daisies would have to use some of their energy to make pigment and therefore would be at a disadvantage compared with unpigmented gray daisies. In such a world no temperature regulation would take place. As they put it, "the gray daisies would cheat."

Stimulated by their criticism I made a model with three species of daisies. All that the new model required was another set of equations to describe the temperature and growth of the gray daisy species. It was a matter of introducing sober-suited middle-management daisies to a world of colorful eccentrics. I charged the dark and light daisies a 1 percent growth-rate tax to make pigments. I am pleased to report that the biologists' cynical view of the world is not supported by this new model, as you

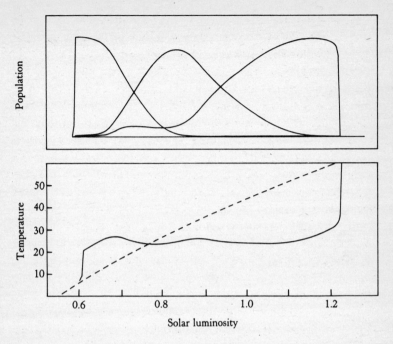

3.2 The evolution of the climate on a three-species Daisyworld with dark, gray, and light daisies present. By comparison, the dashed line in the lower panel represents the temperature evolution in the absence of life.

can see from the results in figure 3.2. Again, the solar luminosity increases from 60 to 140 percent of that of our own Sun. The populations of the daisies are on the upper panel, with dark on the left, gray in the middle, and light on the right. The fact that gray daisies use no energy to make pigment avails them of nothing when their world is too hot or too cold for them to grow. But dark daisies can flourish in the cold, and white ones in the hot. Gray daisies do best when the climate is temperate and when regulation is not needed. In other words, the different species grow when they and their environment are fit for one another.

It would have been sufficient as an answer to the critics merely to have added gray daisies, but having started, I found

that it was almost as easy to make a model that would accommo-
date any number of daisies from one to twenty. I therefore did
this, constructing the model so that, whatever the number of
species, the shade of the daisies varied in constant steps going
from dark to light. Figure 3.3 illustrates the evolution of temperature, daisy population, and other properties of a world with
twenty species of daisies. Like the three-species model, it is of
a world whose star grows hotter as it ages. The lower panel
shows the evolution of the planetary temperature, the middle
panel the evolution of the populations of the different species,
and the upper panel the total biomass and the diversity of the
ecosystem represented in the model. The diversity of the ecosystem is greatest when there is the least stress. When the heat
from the Sun is just right for growth and no effort is required
for temperature regulation, then the greatest number of species
can coexist. When the system is under stress, when it has just
begun to evolve or is about to die, then diversity is least and
the population is almost entirely made up of the darkest or
the lightest species. Indeed, if any species is at an advantage
at such times, it is the darkest or the lightest, not the gray.

But there is much more to this new model than an answer
to the criticisms of skeptical biologists. When I made it I was
ignorant of theoretical ecology, that branch of mathematical
biology that is concerned with interactions among the species
of an ecosystem. As we shall see, Daisyworlds provide an escape
for a science that has been trapped for years by the limitations
of its theories.

In the 1920s, the mathematical biologists Lotka and Volterra
introduced their famous model of the competition between rabbits and foxes. Like Daisyworld it was a simple model, but it
differed in that the environment was taken as infinite and neutral.
The growth of the populations of rabbits or foxes did not affect
the environment, and no environmental changes were allowed
to affect the rabbit and fox populations. The two equations
for this world can be expressed in words as follows: Foxes increase as the numbers of rabbits grow, but rabbits decline as
the foxes increase. In this relationship, there is one stable point

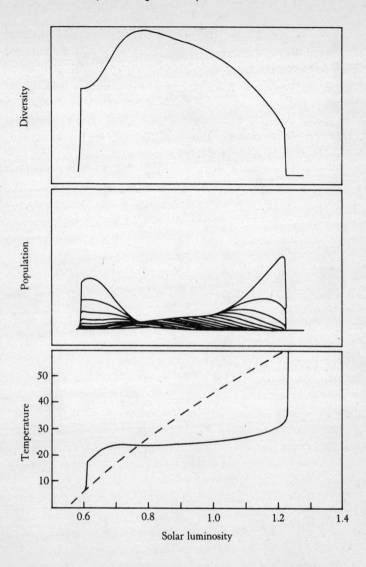

3.3 The evolution of the climate on a 20-species Daisyworld. The lower panel illustrates planetary temperature—the dashed curve with no life present, and the solid line with daisies. The middle panel shows the populations of the 20 different colored daisies, with the darkest appearing first (*left*) and the lightest last (*right*). The upper panel illustrates diversity, seen to be maximum when the system temperature is closest to optimum.

where the two species coexist at a constant ratio. But one bad season that kills off some rabbits, indeed any change in population other than from the model itself, dooms this simple world to cyclical fluctuation from which it can never return to the stable ratio. Note how this happens. If a plague causes a sudden death of rabbits, this will be followed by the death of foxes from starvation. Rabbits breed fast, and soon their numbers are up to and beyond those before the plague. But now foxes begin to increase also, and the rate of rabbit growth slows and then declines as a surfeit of foxes culls them. Soon there are too few rabbits to feed the foxes, the foxes die, and the cycle begins anew.

Does this model world account for observed population swings in Nature? Yes, it does. Field ecologists have shown that population cycles do occur in simple ecosystems, but when we look closer, it seems that the field observations are almost always chosen from diseased or man-made ecosystems where few major species are present and interacting, and when only two of the species are considered (for example, pests attacking the vegetation of an agricultural monoculture, or bacterial disease among plants and animals). In these two-species examples, the populations either cycle periodically or fluctuate in a chaotic and unpredictable manner, and can be successfully modeled by the mathematical successors of Lotka and Volterra's famous fox and rabbit model. What these ecological models and theoretical ecology as a science have so far been unable to explain is the great stability of natural complex ecosystems like the tropical forests or Darwin's tangled bank: "Whereon the wild thyme blew and oxlips and the nodding violet grew."

Ecologists have attempted to overcome the inadequacies of their simple models by including a structured hierarchy of species that are referred to as "food webs." In such a hierarchy there is a pyramid that is surmounted by the top predator, such as a lion, with the smallest numbers. The numbers increase as you go down each "trophic" level, until at the base of the pyramid are the most numerous primary producers, the plants, that provide food for the whole system. In spite of years of effort and computer time, the ecologists have made no real progress towards modeling a complex natural ecosystem such as a tropical rain forest or the three-dimensional ecosystem of the ocean. No models drawn from theoretical ecology can account in mathematical terms for the manifest stability of these vast natural systems.

Indeed, a distinguished ecologist, Robert May, in his book *Theoretical Ecology*, writes in a chapter entitled "Patterns in Multispecies Ecosystems":

> When these kinds of studies are made, a wide variety of mathematical models suggest that as a system becomes more complex, in the sense of more species and a more rich structure of interdependence, it becomes more dynamically fragile. . . . Thus, as a mathematical generality, increasing complexity makes for dynamical fragility rather than robustness.

May goes on to write:

> This is not to say that, in Nature, complex ecosystems need appear less stable than simple ones. A complex system in an environment characterized by a low level of random fluctuation and a simple system in an environment characterized by a high level of random fluctuation can well be equally likely to persist, each having the dynamic stability appropriate to its environment. . . . An important general conclusion is that large and unprecedented perturbations imposed by man are likely to be more traumatic for complex ecosystems than for simple ones. This inverts the naive, if well intentioned, view that "complexity begets stability" and its accompanying moral that we should preserve, or even create, complex systems as buffers against man's importunities. I would argue that the complex natural ecosystems currently under siege in the tropics and subtropics are less able to withstand our battering than are the relatively simple temperate and boreal systems.

This disclaimer recognizes the stability of complex ecosystems in the real world; but the impression remains that diversity is, in general, a disadvantage and that Nature, by disregarding the elegant mathematics of theoretical biology, has somehow cheated.

Obviously, had I known of this work, I would never have attempted anything so foolish as a model with twenty daisies. Fortunately for me I was brought up in that school of science that believes in reading the books after rather than before an experiment. What is it, then, that confers the great stability and freedom from cyclical and chaotic behavior on the Daisyworld models? The answer is that in Daisyworld the species can never grow uncontrollably; if they do, the environment becomes unfavorable and growth is curtailed. Similarly, while daisies live, the physical environment cannot move to unfavorable states; the responsive growth of the appropriately colored daisy prevents it. It is the close coupling of the relationships which constrain both daisy growth and planetary temperature that makes the model behave. Perhaps it is a metaphor for our own experience that the family and society do better when firm, but justly applied, rules exist than they do with unrestricted freedom.

Curious to see if this explanation was correct, I made an additional Daisyworld. In this one, the daisies were grazed by rabbits and the rabbits in turn eaten by foxes—a combination of Lotka and Volterra's model with Daisyworld. To test the stability of this more complex model, I subjected it to periodic catastrophes; on four occasions during the evolution of the model 30 percent of the daisy population was destroyed suddenly as by a plague, and the system then allowed to recover (see figure 3.4). Remarkably, neither the addition of herbivores nor plagues seriously affects the capacity of the daisies to regulate climate. During the normal course of evolution all populations are stable and recover promptly from the perturbations of the plagues. Eventually, the system can no longer cope with the ever-increasing solar output and it fails. As one might expect, the nearer to failure the greater the effect of the perturbations.

The difference between the geophysiological and the ecological view is in the interpretation of perturbation. The geophysiolo-

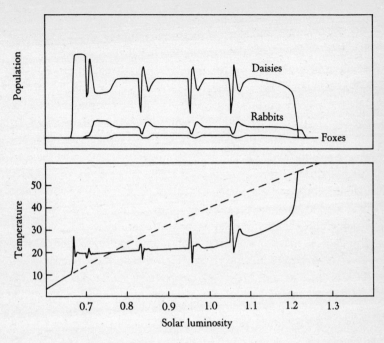

3.4 Daisyworld with rabbits and foxes, perturbed by four plagues that killed 30 percent of the daises.

gists see temperature, rainfall, the supply of nutrients, and so on as variables that might be perturbed. In their view the Gaian system evolved with its physical and chemical environment and is well able to resist changes of this kind. Forests of the humid tropics are normally well watered and shaded by their canopy of clouds; during their existence they are never subjected to prolonged drought as in a desert region. Theoretical ecologists, on the other hand, ignore the physical and chemical environment; to them the environment means the collection of species themselves and a niche is some piece of territory negotiated among the species, rather as one might regard the environment of Switzerland as comprising the people of Italy, France, and Germany. In such a view, perturbations are competition or wars.

The invasion of a tropical forest by humans with chain saws

who would replace it with an agricultural ecosystem is a trau-
matic act. It is like destroying the ecosystem of the model with
twenty species and replacing it with a monoculture of dark
daisies only. Both in Daisyworld and in the forest, such an act
could lead to premature death by overheating, especially if it
took place at a time or place where the Sun was hot. Geophysiologists and ecologists are agreed that the complex systems could
not easily recover from insults like these; where we differ is
over the stability of the monoculture, or the single daisy species.
Geophysiology says that, because these ecosystems are limited
in their ability to interact with the physical environment, they
are unable to sustain their environment when exposed to a
large perturbation. The humid tropics have remained forested,
in spite of changes in the Earth's climate, which would be considered great in human terms but which are trivial on a planetary
scale. The presence of great species diversity assists towards
this robust capacity to withstand climatic change.

In most of the examples of Daisyworld, the Sun has been
shown as steadily increasing its output of heat; an external
perturbation that relentlessly increases in intensity until life
can no longer continue. An alternative way of illustrating the
stability of Daisyworld is to allow life to go on normally at a
constant intensity of sunlight and then suddenly perturb the
world by a change in climate or by some catastrophe such as a
plague or planetesimal impact. Figure 3.5 shows a Daisyworld
with 10 species of colored daisies whose stable existence is suddenly disturbed by a plague that kills 60 percent of the daisies,
regardless of their color. In the lower panel, the dashed line
represents the planetary temperature with no daisies present;
at 40°C it is at the limit for life. The solid line illustrates the
climate with daisies present before, during, and after the perturbation. The temperature stays in the mid-twenties, except when
the plague first hits the daisy populations. When the perturbation
is relaxed the system rapidly restores the status quo. The upper
panel illustrates the variation in the distribution of species before,
during, and after the catastrophe.

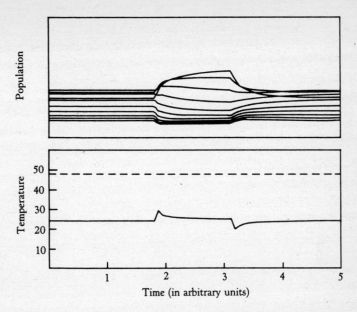

3.5 The effect on climate of a plague that kills off 60 percent of the daisies on a Daisyworld when the solar luminosity is at a constant intensity. Note how homeostasis is restored in both population and temperature during and after the perturbation.

With this perturbation, the system is reluctant to move far from the comfortable state that existed before the change. The most marked effect of the disturbance is in the distribution of the different species of daisies. The rapid response of Daisyworld to change requires positive feedback and involves the explosive growth of those species whose interaction with the climate is the most favorable. This model has 10 species of daisies in a fixed, uniformly distributed range of colors from dark to light. An obvious extension of the research would be to include mutations and the possibility of evolutionary changes in the species. The abrupt change in the distribution of species at the perturbation event and again at its termination indicates the intensity of the selection pressure at these times. The experiment is much in accord with the observations of Stephen Jay Gould and Niles

Eldredge on punctuated evolution. Instead of a steady, gradual change, as in the conventional Darwinian view, there are periods of abrupt and rapid evolution: the punctuation. Gaia theory would expect the evolution of the physical and chemical environment and of the species to proceed always together. There would be long periods of homeostasis with little environmental change or speciation, interrupted by sudden changes in both. These punctuations could be driven internally as a result of the evolution of some powerful species, like humans, whose presence alters the environment, or as the result of external change as from the impact of planetesimals.

The perturbation experiments and the steady-change experiments can be combined into one, as in figure 3.6. Here our

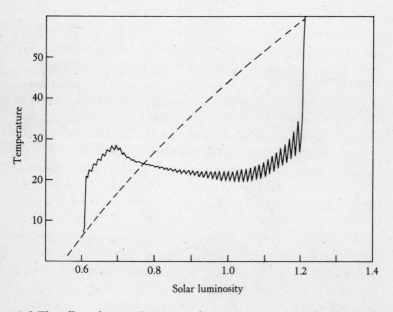

3.6 The effect of a periodic plague of constant intensity on the ability of the daisies to control the climate. Note how the perturbation of the plague is amplified at the times of maximum stress near the beginning and the end of daisy life. The increasing amplitude of the oscillations in the latter part of the curve is reminiscent of the evolution of the present series of glacials and interglacials.

world with 10 species of daisies is evolving as before, but now there is a recurrent plague affecting all colors equally. The plague strikes down 10 percent of the daisies, which are then allowed to recover only to become victims of a new form of the virus. And so the cycle continues throughout the evolution of the model. This experiment graphically illustrates the way that stability, as measured by the capacity to regulate the climate, correlates with diversity. The fluctuations of temperature are greatest at the birth and near the death of Daisyworld when the number of species is least. In the prime of its life, the effect of the perturbations is almost wholly resisted.

We shall come back to this experiment in chapter 6, in connection with the contemporary regulation of temperature by the biota through their capacity to affect the concentration of carbon dioxide in the air. This particular climate control system is nearing the end of its capacity to work and it can be argued that the recent oscillation of climate between ice ages and interglacials is like that near the end of Daisyworld in figure 3.6. The relatively minor perturbations of the Earth's orbital wobble that cause small variations in the amount of heat received from the Sun are amplified by the instability of a moribund system. These arguments do not necessarily apply to glaciations in the more remote past, which are likely to have had different causes.

The Daisyworld models that I have just described are complete but are expressed in ordinary English. Many scientists find such an expression of a theory unsatisfying and prefer the "rigor" of formal mathematical expression. For their benefit, the essence of the Daisyworld model is expressed in the simple diagram, figure 3.7, that my friend and colleague, Andrew Watson, devised. This model is based on a Daisyworld where there are only white daisies present. Because they are lighter in color than the soil in which they grow, they tend to increase the albedo of their locality, which, as a consequence, is cooler than a comparable area of bare ground. Where the daisies cover a substantial proportion of the planetary surface they will influence the mean surface temperature of the planet. The relationship between the area covered by white daisies and the mean surface

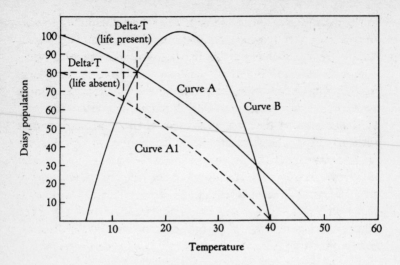

3.7 Regulation by white daisies. The helmet-shaped curve (B) depicts the response of the daisies to temperature, and curves A and A1 are the response of planetary temperature to the area covered by daisies. Curve A1 is for a lower heat input from the star. In the absence of daisies, the change of planetary temperature (Delta-T) would be nearly 15°C whereas in their presence it is only about 3°C.

temperature of the planet is illustrated as curve A. The parallel dashed line (A1) shows how the relationship might be shifted if there were a change in some external variable that influenced the planetary temperature—for instance, if the star that warmed Daisyworld decreased its output of radiant heat.

Like most plant life, daisies grow best over a restricted range of temperatures. The growth peaks near 20°C and falls to zero below 5°C and above 40°C. The relationship between planetary surface temperature and the steady-state population of daisies will be as in curve B, the curve shaped like a helmet. Curves A and B relate the planetary temperature to the population of Daisyworld at steady state; and the steady state of the whole system is specified by the point of intersection of these two curves. In the example, it can be seen that there are two possible steady-state solutions. It turns out that the stable solution is at the intersection where the rates of change of population with temperature are in opposite directions. In mathematical terms, where the derivatives of the two curves have opposite signs. The other intersection does not give a stable solution. If Daisyworld is started at some arbitrary but tolerable temperature it will move to the upper intersection point and settle there.

What happens to this stable state when some change takes place in the external environment? Suppose, for example, that the star warms up as our Sun is said to be doing. If the daisy population were artificially held constant, the planetary temperature will simply follow the change of heat output of the star; there will be a much larger change of temperature than if the daisies were allowed to grow to their new natural steady state, where they would oppose the effect of a change of stellar output.

Very few assumptions are made in this model. It is not necessary to invoke foresight or planning by the daisies. It is merely assumed that the growth of daisies can affect planetary temperature, and vice versa. Note that the mechanism works equally well whatever the direction of the effects. Black daisies would have done as well. All that is required is that the albedo when the daisies are present be different from that of the bare ground. The assumption that the growth of daisies is restricted to a narrow range of temperatures is crucial to the working of the mechanism, but all mainstream life is observed to be limited within this same narrow range. The peaked growth curve (B) is common to other variables besides temperature, for example pH (it can be too acid or too alkaline: neutrality is preferred). Similar restrictions apply to most nutrients; too much is poisonous, too little causes starvation.

The choice of a parabolic relationship for the response of daisy growth to temperature is arbitrary, and some people suggested that in real life the relationship takes on different or more complex forms. To test this objection, the Daisyworld model was run with different relationships between growth and temperature. These all retained the 5 to 40°C limit but ranged in shape from rectangular (a constant growth at all temperatures) to triangular (a linear increase up to a maximum, and then a linear decrease back again). Semicircular and rhombic relationships were also tried. The only limitation observed was that the models became unstable when a horizontal section was present in the relationship. Look back at figure 3.7 and imagine the helmet-shaped curve is replaced by one shaped as a top hat. The horizontal section, the top of the hat, would imply constant growth at any temperature within the 5 to 40°C range; consequently no regulation of temperature could take place. Such models are very similar to the simple competitive models of population biology where the environment is ignored and which are notoriously unstable. So long as the rate of growth of daisies varies with their temperature up to a maximum and then declines, even the gentle curvature of the top of a semicircle provides a working model.

In Nature, the shape of the relationship linking growth with some environmental variable frequently comes from the combina- tion of a logarithmic rise overtaken by a logarithmic decay. In chapters 5 and 6 this relationship forms part of the model for the regulation of atmospheric oxygen. As oxygen increases in abundance the growth of consumers increases, but oxygen in excess is poisonous. Too little oxygen and too much both are bad; there is a desirable sufficiency.

Daisyworld differs profoundly from previous attempts to model the species or the Earth. It is a model more like those of control theory, or cybernetics, as it is otherwise called. Such models are concerned with self-regulating systems; engineers and physiologists use them to design automatic pilots for aircraft or to understand the regulation of breathing in animals, and they know that the parts of the system must be closely coupled if it is to work. In their parlance, Daisyworld is a closed-loop model. Devices that are not self-regulating are often unstable. Engineers refer to them as "open-loop"; the loops are the feedback links between the parts of the system. Daisyworld is not identical in form to an engineered device; a key difference is the absence, in Daisyworld, and perhaps in Gaia also, of "set points." In manufactured systems, the user sets the temperature, the speed, the pressure, or any other variable. The value chosen is the set point, and the goal of the system is to keep that value, however the external environment changes. Daisyworld does not have any clearly established goal like a set point; it just settles down, like a cat, to a comfortable position and resists attempts to dislodge it.

Because of the tribalism that isolates the denizens of the scientific disciplines, biologists who made models of the competitive growth of species chose to ignore the physical and chemical environment. Geochemists who made models of the cycles of the elements, and geophysicists who modeled the climate, chose to ignore the dynamic interactions of the species. As a result, their models, no matter how detailed, are incomplete. It is as if, in figure 3.7, the biology of the relationship between daisy population and temperature was considered without reference to the complementary geophysics of the relationship between temperature and daisy population. An engineer or physiologist would instantly recognize that such an approach was "open-loop," and consequently of little value except as an example of an extreme or pathological condition. There is also something pathological about the hubris of those scientists who boast of their special knowledge and of their disdain for that of other scientific disciplines. Sixty years ago that wise and generous-minded American theoretical biologist, Alfred Lotka, described the model of the competition of rabbits and foxes in *The Elements of Physical Biology*. It has been the inspiration of countless researchers in population biology ever since, yet none can have heeded his warning on page 16:

This fact deserves emphasis. It is customary to discuss the "evolution of a species of organisms." As we proceed we shall see many reasons why we should constantly take in view the evolution, as a whole, of the system (organism plus environment). It may appear at first sight as if this should prove a more complicated problem than the consideration of a part only of the system. But it will become apparent, as we proceed, that the physical laws governing evolution in all probability take on a simpler form when referred to the system as a whole than to any part thereof.

For three generations since, theoretical ecologists have modeled the evolution of ecosystems while ignoring the physical environment; and three generations of biogeochemists have modeled the cycles of the elements without ever including the organisms as part of a dynamic and responsive system. In Alfred Lotka's time the resolution of systems of nonlinear differential equations, even for a simple model like Daisyworld, was a daunting task. With the ready availability of computers there is now no need to continue with models constrained by the narrow limits of a single scientific discipline.

Lotka's insight, that modeling would be simpler with the whole system than with any part of it, is amply confirmed by modern mathematics. Systems of equations of the type describing model systems in theoretical ecology and biogeochemistry are notorious for their chaotic behavior, so much so that they now seem more interesting as playthings, or a new form of graphic illustration. The mathematics of natural phenomena, when constrained within a single discipline, can become so intricate and complex that world after world of colorful abstraction opens at every new level investigated. It is small wonder that practitioners of the various disciplines imagine that in these imaginary worlds they see glimpses of real world whereas in fact they are lost in the fractal dimensional world of a Mandelbrot set that goes on forever at every level from minus to plus infinity. The delusion is encouraged by professional mathematicians who find similarities between their mathematical theories and the pathologies of the real world, and the numerous modern mathematical scientists whose contemplation of the demons of hyperspace—the "strange attractors" of chaos—is much more beguiling than the dull old real world of Nature.

Daisyworld provides a plausible explanation of how Gaia works and why foresight and planning are not required for planetary regulation. But what evidence is there of practical mechanisms? What predictions from Gaia theory have been confirmed? Is it testable? The first test was the Viking mission to Mars. That expedition confirmed the prediction, from atmospheric analyses by infrared astronomy, that Mars was lifeless. Michael Whitfield, Andrew Watson, and I predicted that the long-term regulation of carbon dioxide and climate is by the biological control of rock weathering. These and other tests are described in the three chapters that follow. It matters little whether Gaia theory is right or wrong; already it is providing a new and more productive view of the Earth and the other planets. Gaia theory provokes a view of the Earth where

1. Life is a planetary-scale phenomenon. On this scale it is near immortal and has no need to reproduce.

2. There can be no partial occupation of a planet by living organisms. It would be as impermanent as half an animal. The presence of sufficient living organisms on a planet is needed for the regulation of the environment. Where there is incomplete occupation, the ineluctable forces of physical and chemical evolution would soon render it uninhabitable.

3. Our interpretation of Darwin's great vision is altered. Gaia draws attention to the fallibility of the concept of adaptation. It is no longer sufficient to say that "organisms better adapted than others are more likely to leave offspring." It is necessary to add that the growth of an organism affects its physical and chemical environment; the evolution of the species and the evolution of the rocks, therefore, are tightly coupled as a single, indivisible process.

4. Theoretical ecology is enlarged. By taking the species and their physical environment together as a single system, we can, for the first time, build ecological models that are mathematically stable and yet include large numbers of competing species. In these models increased diversity among the species leads to better regulation.

We have at last a reason for our instinctive anger over the heedless deletion of species; an answer to those who say it is mere sentimentality. No longer do we have to justify the preservation of the rich variety of species in natural ecosystems, like those of the humid tropical forests, on the feeble humanist grounds that they might, for example, carry plants with drugs that could cure human disease. Gaia theory makes us wonder if they offer much more than this. Through their capacity to evaporate vast volumes of water vapor through the surface of their leaves, trees serve to keep the ecosystems of the humid tropics and the planet cool by providing a sunshade of white reflecting clouds. Their replacement by cropland could precipitate a regional disaster with global consequences.

4

The Archean

In the beginning there was nothing, not even space or time.
JOHN GRIBBIN, *Genesis*

Life began a long time ago. The date of the event is not known, but it was at least three thousand six hundred million years before we were born. Numbers as large as this are anesthetic and paralyze the imagination. A different scale of reckoning time is needed to reach back to those bacteria, our ultimate grandparents. In science, the usual way of taming outrageous numbers is to express them as powers of ten. Make every step ten times larger or smaller than the one before. In his book *Timescale: An Atlas of the Fourth Dimension,* Nigel Calder illustrates the Earth's history in this way. He reminds us how easily this logarithmic sense of time can prevent us from recognizing how long life has occupied the Earth; to say that life began 3.6×10^9 years ago does not help. On a linear scale of measure, the origin of life was about a thousand times more remote than the origin of humans. In this book, I shall use a scale of eons, which represent a thousand million years. Life started at least

3.6 eons ago, during the period geologists call the Archean, the period that runs from the Earth's assembly 4.5 eons ago to 2.5 eons ago when oxygen first dominated the chemistry of the atmosphere.

Gaia is as old as life; indeed, if the Big Bang that started the Universe was 15 eons back from now, she is a quarter as old as time itself. She is so old that her birth was in the region of time where ignorance is an ocean and the territory of knowledge is limited to small islands, whose possession gives a spurious sense of certainty. In this chapter I invite you to join with me in speculating about the infant Gaia and the problems she faced in taking on her inheritance, the Earth. When we look at the Archean period in the light of Gaia theory, we see a planet radically different from that depicted by the conventional wisdom of present-day science. It is a planet where life does not just adapt to the Earth it finds itself upon, but also adapts the Earth to make it and keep it a home.

The best way to illustrate the powerful presence of Gaia is to consider what the Earth would be like without life. It will be argued that the present-day Earth could be an arid place like Mars or Venus had life not appeared upon it. We cannot make such a comparison for the Archean because we know so little about the Earth then. What we must do, therefore, is make a best guess about the condition of the Earth before life, and then consider the changes there would be when life took charge. By asking what the Earth was like before life began, we are in a way hanging up a neutral back cloth before which can be clearly seen the colorful changes made by life.

The trouble with doing this is that the back cloth is so old as to have all but moldered away. Looking back in time is like using a telescope to view the limits of the Universe. We can see faintly luminous objects. Astronomers make a convincing case that the distance of these objects is so great that the light now seen started its journey to the Earth 3.8 eons ago. This is close to the time geologists believe the first bacterial cells came into existence. They are probably correct, but the only certainty about such remote times and places comes from the great second law of thermodynamics. Enigmatically, it states that the beginning and the end of the Universe are unknowable. As time and distance lengthen, the once fair face of knowledge grows pockmarked with craters of ignorance. In the end, the features can no longer be recognized.

Information theory teaches that, in the presence of a constant amount of noise, the power required to send a signal across a gulf of space and time increases exponentially with the distance to be traveled. In simpler words: as the distance or the time lengthens, vastly more power is needed to transmit the same message. The happenings on Earth a mere 5000 years ago are far from known with certainty. Just imagine how large a signal would be needed to transmit information about the beginning of the Universe 15 eons ago. This may be why the Big Bang theory that the Universe began by the explosion of a primeval particle is inevitable. Nothing short of the explosion of the Universe itself could send a signal from so long ago. All that now lingers is the faint rumble of the cosmic microwave background radiation. But all other theories of the origin are without evidence.

There is a clever way to gather information about events as ancient as the start of life that avoids the otherwise universal tendency of messages to age and die. It comes from the nearly miraculous property of living matter to overcome the attenuating tendency of time. Not only has Gaia stayed alive from the beginning; she has also provided a noise-free channel of chemical messages about those ancient times.

If you stand on a hilltop and shout, you will not be heard more than a mile away. If you use a loud-speaker, you might be able to send a message 5 miles. Even exploding an H-bomb would make your point only for a few hundred miles. The alternative is to tell a friend who will take the message and pass it on by word of mouth. By this means, the message could travel without difficulty to the ends of the Earth. In a similar way, living organisms pass on the programs of the cell from one generation to the next. There is every reason to believe that we share with the first ancient bacteria a common chemistry, and that the natural restrictions on the existence of those ancient bacteria tells us what the environment of the early Earth was like. By transmitting coded messages in the genetic material of living cells, life acts as a repeater, with each generation restoring and renewing the message of the specifications of the chemistry of the early Earth. It is a much better channel of information than the record of the rocks. It is precise, but unfortunately it is inaccurate in the way that a message passed by word of mouth is precise and makes sense, but inevitably "mutates." There is the wartime joke that hides a truth: how the message passed by word of mouth, "Send reinforcements, we are going to advance" mutated into "Send three and four pence, we are going to a dance." If we wish to know life's origins from genetic information we need to be prepared to reconstruct the truth from errors of this kind.

By contrast, most of the geological information about the early Earth came from another big bang. It had to be large to send a signal so far. It was the explosion of a star-sized nuclear bomb, a supernova. We tend to ignore that we oddities, who use combustion as a source of energy, inhabit a nuclear-powered Universe. The power plants, the stars, run for billions of years with utmost reliability. But just as the most dependable systems we design can still have the occasional accident, so some kinds of stars occasionally explode. Fortunately for us, one of them did and gave us the start we needed. Fortunately, also, our Sun is not of the exploding kind; it is neither big enough nor old enough.

How can we be so sure that the Earth's origin was connected with the explosion of a supernova? We are sure because, even today, the Earth is radioactive, and also because the Earth is made of elements like iron and silicon and oxygen that cannot be made in the normal processes of stellar evolution. In the Sun and similar stars, hydrogen is fused to generate helium, and the reaction liberates the vast outpouring of heat that keeps us warm even 100 million miles away. But no ordinary fusion process can make elements such as iron, nor those such as uranium, which are heavier. It takes energy to make such elements.

Powering a star by fusing iron to make uranium is like trying to burn ice in a furnace. This is not the place for fine details of element synthesis in exploding stars, except to say that in one kind of explosion the key part of the event is the gravitational collapse of the star. The innermost regions support the fantastic pressure of all the mass of the star trying to fall in. In their active life, the heat generated by nuclear reactions at the center of the star sustains a pressure high enough to balance the inward force of gravitation. It is just like a space rocket at the moment of takeoff; the weight of the vehicle is supported by the blast of flame. But the outer layers of the star cannot escape the pull of gravitation and, when the fuel runs out, it collapses. It is then that the heavy elements are synthesized. Some proportion of them is violently ejected as the outer and still unburnt layers of the star explode.

We still do not know how the Solar System and the Earth came together as a result of that supernova; nor how its radioactive debris became so large a part of our planet. But radioactivity is a marvelously accurate clock, and has precisely ticked away the time since that explosion 4.55 eons ago. We are so used to thinking of radioactivity as artificial that we easily ignore the fact that we ourselves are naturally radioactive. Every minute, in each one of us, a few million potassium atoms undergo radioactive decay. The energy that powers these minuscule explosive atomic events has been locked up in potassium atoms ever since that stellar explosion long ago. The element potassium is radioactive but it is also essential for life. If it were removed and replaced by the very similar element, sodium, we should die instantly. Potassium, like uranium and thorium and radium, is a long-lived radioactive nuclear waste of the supernova bomb. When potassium atoms decay, they are transmuted to form atoms of calcium and of the noble gas argon. The one percent of argon that goes to make up the atmosphere has, over the course of the Earth's history, mostly come from potassium in this way. In the rocks, the radioactive elements uranium and thorium are present at several parts per million. Their rate of decay is so slow that most of what was originally present still remains, except for the uranium isotope ^{235}U, nearly all of which has decayed. It is the heat generated by the decay of these radioactive elements that keeps the Earth's interior hot and drives the movements of the crust.

The evidence from the rocks suggests that life began between 0.6 and 1 eon after the Earth had come together as a recognizable planetary body. The evidence is a difference in the proportions of the atoms of the stable element carbon. Carbon atoms exist on Earth in three forms: the common form weighs 12 atomic units, but there is a proportion weighing 13 units and a small trace of radioactive carbon weighing 14 units. These different-weight atoms are called isotopes. The proportion of the 12 and 13 isotopes, in the carbon of rocks made in the absence of life, is recognizably different from the proportion in carbon from rocks that were once living matter; this is because the chemistry of living matter segregates the isotopes. By measuring the isotopic composition of ancient rocks it is possible to distinguish those that were made when life was present from those that formed before life began. The most certain pre-life rocks we have come not from the Earth but from the Moon or from meteorites. These are as old as 4.55 eons. The isotopic proportion of these dead-matter rocks is easily distinguished from those laid down on Earth 3.6 eons ago. The oldest sedimentary rocks on Earth so far found are 3.8 eons old, and they come from a place called Isua in Greenland. I recall the German geochemist Manfred Schidlowski describing these ancient rocks in a 1973 lecture, and speculating that the carbon atoms within them when they formed showed an isotopic distribution suggestive of the presence of life.

The period before life has left no rocks from which we could reconstruct the details of the environment in which they formed; 4 eons or more of weathering and grinding has erased the record. It is likely to have been a time of unimaginable violence, with small planets left over from the condensation of the Solar System still crashing in. (The impact of a planetesimal a mere 6 miles in diameter can leave a crater 200 miles across, and splash molten rock and gas far out into space.) It left an Earth as cratered as the Moon. It was a period well named the Hadean.

The chemistry and physics of the period just before life began can only be surmised, and it will be interesting to watch as speculations blossom about the amazing and turbulent history of the Earth's beginnings. You can see, however, the difficulty in weaving the neutral back cloth referred to earlier. Therefore we will have to make the best we can of the information available, starting with the atmosphere.

The atmosphere is the face of the planet, and it tells, just as do our faces, its state of health and even if it is alive or dead. As we saw in chapter 1, planetary life is obliged to use any mobile media—that is, the air or the oceans—as conveyors of raw materials and as conduits for waste products. Such a use of these fluid media leads to profound changes in their chemical composition and to their departure from the near-equilibrium steady state characteristic of a nonliving planet. Dian Hitchcock and I used the absence of such changes in the atmosphere of Mars and Venus as evidence for the absence of life long before the Viking and Venera landers looked for and failed to find it. These dead planets are visually as well as chemically a neutral background against which the living planet Earth shines like a dappled sapphire.

There are many reasons why the atmosphere is so much more revealing about life than are the ocean or the crustal rocks. It is the region of rapid chemical change under the influence of sunlight; no mixture of gases capable of chemical reaction can long remain unchanged in the atmosphere. If we find a combustible gas like methane present with oxygen in a sunlit atmosphere, we know for certain that something is constantly making them both. No such conclusion could be drawn about air in a sealed underground cave. It is the sunlight that constantly keeps ignited all possible chemical combustions. Then the atmosphere has the smallest mass of all the compartments that life encounters; apart from the small concentration of rare gases like argon and helium, all other gases of the air have recently existed as part of the solids and liquids of living cells. The atmosphere also has an immediate effect on the climate and chemical state of the Earth, features of fundamental importance to life. A similar exchange takes place between life and the oceans and the rocks, but it is much slower in pace and the cycles of life are diluted by materials used long ago but now discarded.

The Earth, just before it became the habitat of life, then, must have been a dead planet whose atmosphere was near to equilibrium. At this time just before life, before Gaia, the atmosphere would have been in what scientists call the "abiological steady state." This wordy phrase is to distinguish the real planet—which has hurricanes and tornadoes, volcanoes and whirlpools—from the fiction of the utter stillness of an equilibrium planet.

The early Earth is thought to have had on its surface the chemical components from which life assembled, chemical compounds that are called "organic"—such as amino acids, the subunits of protein; nucleosides, the subunits of the molecules of our cells that carry their genetic information; sugars, the subunits of polysaccharides; and many other essential parts waiting for the final act of assembly. It is important to recognize that these chemicals, although we regard them as characteristic of life, are also the products of the abiological steady state. The mere presence of such compounds on an oxygen-free planet is not by itself evidence for life. It is evidence only of the possibility of its formation.

Not only was the Earth's chemistry just right for life to start, the climate also must have been favorable. Some ancient rocks show evidence of having been formed by the sedimentation of particles. Their layered structure suggests an origin in a shallow lake or sea and, therefore, of the presence of free water. The existence of life and pre-life chemicals requires a temperature range between 0 and 50°C. The Earth could not have been frozen, nor could it have been hot enough for the seas to boil.

In an important paper in 1979, three atmospheric chemists and climatologists, T. Owen, R. D. Cess, and V. Ramanathan, reported calculations to determine the average temperature of the Earth at the time life began. They used the general consensus of astrophysicists, that stars grow hotter as they age, and supposed that the output of heat from the Sun was 25 percent less than it is now. They took values for the approximate amount of carbon dioxide gas that had escaped (or outgassed) from the Earth's interior. From this, they were able to calculate that the mean surface temperature of the Earth was 23°C; typical of the tropics today. Their calculations required the presence of 200 to 1000 times as much carbon dioxide in the air as there is now. Much would depend on the quantity of nitrogen present. If then, as now, nitrogen was the principal atmospheric gas, then the lower pressure of carbon dioxide would have sufficed. Also important, according to my friend the climatologist Ann Henderson-Sellers, would have been the distribution of water as oceans, snow, ice, clouds, and water vapor. Not surprisingly there is still debate about the climate on the occasion of life's start. Calculations by the climatologist R. J. Dickinson in 1987 suggest it may have been a few degrees cooler, in other words just about the same as now.

The idea was that the lack of warmth of a cooler Sun could have been offset by a blanket of "greenhouse" gas. Gases with more than two atoms in their molecules have the interesting property of absorbing the radiant warmth, the infrared radiation, that escapes from the Earth's surface. These gases, which include carbon dioxide, water vapor, and ammonia, are transparent to the visible and the almost visible infrared radiation. These are the parts of the Sun's spectrum that carry most of its energy; radiant heat in this form will penetrate the air and warm the surface. The same gases are opaque to the longer wavelength infrared that radiates from the Earth's surface and lower atmosphere. The trapping of the warmth, which otherwise would escape to space, is the "greenhouse effect"; so called because it is like, although not the same as, the warming effect of the glass panes of a greenhouse. The first proposal that a gaseous greenhouse warmed the Earth was made by a distinguished Swedish chemist, Svante Arrhenius, in the last century.

H. D. Holland, in *The Chemical Evolution of the Atmosphere and the Oceans,* gives a clear and readable statement of the probable state of the Earth just before Gaia awoke. In summary, he proposes an Earth with an atmosphere rich in carbon dioxide, with nitrogen present but bereft of oxygen, and with traces of gases such as hydrogen sulfide and hydrogen present. The oceans were laden with iron and other elements and compounds that can only exist in solution in the absence of oxygen. Among these could have been reduced compounds of sulfur and nitrogen. The presence of these gases and substances is important, because they are reducing agents—they readily react with, and so remove, oxygen. Such an Earth would have a vast capacity to absorb oxygen and prevent its appearance in the free state. This proposal seems so reasonable that I shall take it as if it were a fact and use it as a key to understanding the evolution of the Archean period of the Earth's history.

One other condition of the nascent Gaia is that three times as much internal heat was produced as now. This was because the Earth was more radioactive; less time had elapsed since the supernova that made it, and the fallout was still hot. It would be wrong, though, to think that this internal heat had an appreciable effect on the surface temperature of the Earth. The heat flux from below was trivial compared with that received from the Sun. The principal effect of greater production of internal heat would have been more vigorous volcanism, a higher output of gas to the air, and a more rapid reaction of volcanic rocks with the ocean waters. One of these reactions, that between the ferrous iron of basalt rock and water, can produce hydrogen gas. The continuous production of hydrogen would have had two important consequences. First, the maintenance of an oxygen-free atmosphere and surface favorable for life chemicals to accumulate. Second, the loss of hydrogen to space. The Earth's gravitational field is not strong enough to hold down the light atoms of hydrogen. If hydrogen escape had continued, we might have lost much of the oceans or even arrived at the arid state of Mars and Venus. (Such an escape cannot take place now because hydrogen would react biochemically in the oceans and with the abundant oxygen in the atmosphere to form water.

Although it carries two hydrogen atoms, water is too heavy a molecule to escape directly into space. Another restraint on the direct loss of water from Earth is its tendency to freeze out and fall back as ice crystals from frigid regions of the air.)

That, then, was the Earth before life. We can accept as reasonable the view that life started from the molecular chemical equivalent of eddies and whirlpools. The power that drove them was the flux of energy from the Sun and also the free energy of a hot young Earth. Prigogine and Eigen have plausibly formu-lated the physical mechanisms by which chemicals and cyclical reactions come together as dissipative structures of protolife. The stepwise evolution from protolife to the first living cell by a process of natural selection does not seem to me so difficult an intellectual pill to swallow. It would be interesting to know if protolife was tightly coupled to its environment and had the capacity to regulate. Two geochemists, A. G. Cairns-Smith and Leila Coyne, have alternatively suggested that the solids of the environment played a crucial part in life's origin. To my mind their ideas help to crystallize the supersaturated arguments, even though their details are disputed. The problem with dissipative structures of the fluid state is that they dissipate too soon. If they are to evolve to more permanent structures, something solid is needed to serve as an anchor or to house them. Again, the mental image of a wind instrument like a flute is helpful in this otherwise confusing topic. Just blowing makes a hiss of unruly dissipating eddies. But when the flutist blows across the port hole of the flute, the eddies are caught and tamed within the solid bounds of its hollow resonant tube to emerge as coherent musical notes. In their evolution, living organisms, too, seem to have used the security of the solid state of matter to store and pass on to their descendants the message of existence. The special solid state of the aperiodic crystals of DNA store the programs of the cell, and give organisms a span far beyond that of a dissipating eddy or a chemical cycle.

The first living cells may have used as food the abundant organic chemicals lying around; also the dead bodies of the less successful competitors and the bodies of the successful ones that died of natural causes. These supplies of raw material and energy may soon have become scarce, and at some early time organisms discovered how to tap the abundant and inexhaustible energy of sunlight to make their own food. It is thought that the first of the photosynthesizers used the less demanding photochemical dissociation of hydrogen sulfide. Soon the real prize, how to use light energy to break the strong bonds binding oxygen to hydrogen and carbon, was won. Bacteria now called cyanobacteria, because of their blue-green color, did just this and are the predecessors of all green plants that now exist.

There was a complete planetary system in the Archean. At the surface—in the sunlight—there were the primary producers, cyanobacteria (ancestors of those shown in figure 4.1), that used solar energy to make organic compounds and replicate themselves. They also would have made oxygen, but the abundance of reactive inorganic chemicals in their environment would have kept this gas close to the site of its production. Also present in the early ecosystem were the methanogens that gained material and some energy by rearranging the molecular products of the producers. The presence of these "scavenger" organisms would have assured the continuous disposal of the products and corpses of the photosynthesizers and the return to the environment of the essential element carbon as methane and carbon dioxide. They could not, as we and animals do, eat the cyanobacteria and use the food they had synthesized; to do this they would have needed oxygen.

I suspect that the origin of Gaia was separate from the origin of life. Gaia did not awaken until bacteria had already colonized most of the planet. Once awake, planetary life would assiduously and incessantly resist changes that might be adverse and act so to keep the planet fit for life. Sparse life hanging on in oases could never have the power to regulate or oppose the unfavorable changes that are inevitable on a lifeless planet. Sparse life would only be found at the birth or death of a Gaian system.

The successful evolution of the photosynthesizers could have led to the first environmental crisis on Earth, and I like to think the first evidence of Gaia's awakening. In gaining their

4.1 Photomicrographs of cyanobacteria. These are the organisms that first used the energy of sunlight to produce organic materials and oxygen. They have been, both in the free state, and as endosymbionts, the primary producers from the beginning of the Archean until now. (Photographs courtesy of Michael Enzien.)

energy, the photosynthesizers would have used the carbon dioxide of the air and the oceans as their source of carbon. Just as we have a carbon dioxide problem now, so might they. We are beginning to realize that the benefits of burning fossil fuel as a source of energy are offset by the dangers inherent in the accumulation of carbon dioxide; it could lead to overheating. The danger faced by the photosynthesizers was the reverse. The cyanobacteria used the carbon dioxide as food. They were eating the blanket that kept the Earth warm. There was at that time a vigorous input of carbon dioxide from volcanoes, but the potential capacity of the bacterial sink could have far exceeded this source. If there had been only photosynthesizers, their abundant bloom over the oceans and on the surface could have reduced the carbon dioxide in a few million years to dangerously low levels. Long before the cyanobacteria ran out of carbon dioxide to eat, the Earth would have cooled to a frozen state and life could have persisted only where heat from below could melt the ice, or moved into a cycle of freezing and thawing as carbon dioxide from volcanoes accumulated and was then removed again. I think that neither of these calamities ever happened. The persistent presence of sedimentary rocks from 3.8 eons ago until now suggests that liquid water has always been present and the Earth has never been entirely frozen. What I would like to propose is a dynamic interaction between the early photosynthesizers, the organisms that processed their products, and the planetary environment. From this there evolved a stable self-regulating system, a system that kept the Earth's temperature constant and comfortable for life.

Before venturing further into this imaginary reconstruction of life with Gaia in the Archean, I must emphasize that it will be no more than a flight of fancy. Solid evidence from the early Archean is scarce, and many different models can be made of it. The eminent geologist, Robert Garrels, often reminds me that in his model of the early Earth, the carbon dioxide remained abundant (about 20 percent by volume) and the Earth was hot (40°C or higher). The point of my model is not to argue for one or other global Archean ecosystem, but rather to illustrate how Gaia theory provides a different set of rules for planet models. The possible climatologies and geologies of a living planet are wholly different from those of a dead planet bearing life as a mere passenger. Having said this, let us continue with our "let's pretend."

In the Archean, photosynthesizers used carbon dioxide and converted it to organic matter and oxygen or its equivalent; just as plants do today. The oxygen would have been mopped up immediately by the ubiquitous oxidizable matter of the environment; the iron and sulfur in the oceans. There was no significant population of oxidizing consumers grazing the photosynthesizers and returning carbon to the environment as carbon dioxide. There was, except in juxtaposition with the producers, no oxygen for consumers to breathe. Instead, there were the methanogens, scavengers and descendants of the original decomposers of organic chemicals. These early bacteria, capable of existence only in the absence of oxygen, lived by decomposing organic matter and converting the carbon in it to carbon dioxide and methane which they return to the air. They served in the Archean, like the consumers of today, to return to the air almost as much carbon as had been removed by the photosynthesizers.

But what of the methane? Methane is a greenhouse gas like carbon dioxide, but it is much less stable in the atmosphere; it decomposes in solar ultraviolet light and reacts with hydroxyl radicals—small molecules, with one atom each of hydrogen and oxygen, that are amazingly reactive and scavenge from the air all but the most stable molecules. It is reasonable to suppose that, in the Archean, this photochemical reaction zone would have been high in the atmosphere but at a level where the air was still dense enough to absorb ultraviolet. When ultraviolet breaks down methane, the products combine and recombine with other molecules to form a suite of complex organic chemicals. Suspended high in the stratosphere, these products could include droplets and particles; an upper-atmospheric smog. Such a layer could have profoundly changed the Archean environment. In its presence, the ultraviolet and visible radiation from the Sun would have been absorbed, and the region where the absorption occurred would have grown warmer. The presence of this warm layer in the atmosphere placed an "inversion" lid on the lower atmosphere, and would have reversed the normal tendency for a fall in temperature as one ascends from the surface. In other words, methane smog would have been the Archean equivalent of the ozone layer and would have acted, just as ozone does, both to stabilize the existence of the stratosphere and to filter out ultraviolet radiation.

The existence of a lid, the "tropopause," above the lower atmosphere would have reduced the flux of methane to the regions where it was destroyed by ultraviolet; just as the foul air is trapped beneath the inversion layer of this century's air-pollution smogs. By this means, the methane concentration could have built up sufficiently to be useful as a greenhouse gas; also, its reaction products in the stratosphere, including water vapor, would have served in the same way. The screening out of ultraviolet by the smog layer would have protected other unstable gases such as ammonia and hydrogen sulfide and allowed them to accumulate to some extent in the lower atmosphere. Ultraviolet normally decomposes hydrogen sulfide and other similar gases, both directly and by other photochemical reactions that produce the hydroxyl radicals. It is conceivable that the lower atmosphere, shielded by the methane smog, contained some free oxygen coexisting with an excess of methane; just as there is free methane in small quantities coexisting in an excess of oxygen in the air we breathe now. This would be even more probable if the photosynthesizers existed in self-contained communities at the surface. Some of the oxygen they made would then diffuse into the air and persist for much longer than that released into the oxygen-hungry waters of the oceans. In a fully detailed model, we ought to include gases such as nitrous oxide, carbonyl sulfide, and methyl chloride; all are components of our present atmosphere. For this model, it is enough to bear in mind this possibility and the amazing and intricate series of reactions and consequences that could come from their presence.

How stable would a planetary ecosystem be that was made up from photosynthesizers using carbon dioxide and decomposers that converted organic matter back to carbon dioxide and methane? In many ways the photosynthesizers are like white daisies; their growth cools the Earth by removing carbon dioxide. The methanogen decomposers are like dark daisies; their growth makes for warmth by adding greenhouse gases to the air. It is not difficult to model the simple world I have just described, constructed just as were the daisy models of chapters 2 and 3. Figure 4.2 illustrates the time course of the evolution of the Earth's average temperature, the atmospheric gases, and of the population of the bacterial ecosystem. The model used H. D. Holland's estimate of the input of carbon dioxide from volcanoes, but the sink for carbon dioxide, by the weathering of rocks, was assumed to increase as the ecosystem developed. I based the climate regulation mainly on the capacity of carbon dioxide and methane to act as greenhouse gases. A small additional effect was assumed to occur—the colonization of the land surfaces would increase cloudiness and tend to increase the back reflection of sunlight.

The upper part of figure 4.2 illustrates the time history of the temperature of this anoxic world with and without the presence of life. The dashed line is the expected temperature rise of a lifeless planet that has enough carbon dioxide to make up an atmospheric pressure of 100 millibars; about one-tenth of the present total atmospheric pressure. The bulk of the atmosphere was assumed to be nitrogen as it is now on Earth. The star was assumed to be 25 to 30 percent less luminous than the Sun is now, but to warm up as time passed in the same way as did the Sun. The solid line marks the temperature of the model world where photosynthesizers are coexisting with methanogens. Note the abrupt and sudden fall in temperature from around 28°C to 15°C after life starts. This is due to rapid decline in the abundance of the greenhouse gas, carbon dioxide, as the photosynthesizers use it to build their bodies. The fall does not continue until the planet freezes because the new greenhouse gas, methane, and some carbon dioxide are returned to the air by the methanogens. Once a steady state is established, this simple cybernetic system regulates the planetary temperature

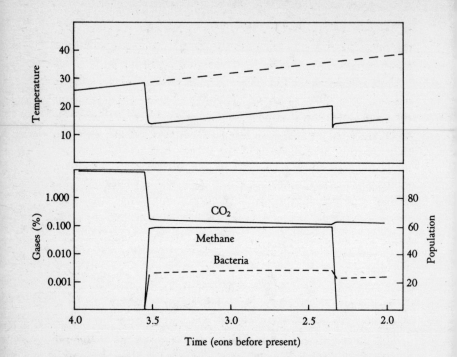

4.2 Model of the Archean before and after life. The upper panel shows the climate with and without life and the lower panel the abundance of the atmospheric gases and bacterial population as the system evolved. The scale for the abundance of atmospheric gases is logarithmic; the scale for population is in arbitrary units.

throughout the Archean. The sudden fall in temperature at about 2.3 eons ago marks the end of the Archean in the model and the appearance of an excess of free oxygen in the air. This event would have led to a decline of methane gas to near its contemporary abundance, thereby removing its greenhouse effect. The model matches the Earth's ancient history. There is no evidence of unusual temperature change during the Archean, and there was a cold glacial period 2.3 eons ago that may have coincided with the appearance of atmospheric oxygen. The lower part of figure 4.2 shows how the total population of bacteria and the abundance of carbon dioxide and methane changed as the model evolved. The start of life is seen to coincide with the fall of carbon dioxide and the rise of methane. The end of the Archean is marked by the disappearance of methane.

This simple model, like Daisyworld, is robust and is not easily disturbed by changes in solar input, bacterial population, or the input of carbon dioxide from volcanic sources. It is sensitive to changes in the range or form of the relationship between the growth of the bacteria and the temperature of their environment. The model is based on the assumption that growth of the bacterial ecosystem ceased at freezing point, was maximum at 25°C, and ceased again at temperatures above 50°C. Like Daisyworld, there is an abrupt change of conditions when life starts. Living organisms grow rapidly until a steady state is reached where growth and decay are in balance. This rapid, almost explosive, tendency to expand to fill an environmental niche acts as an amplifier. The system moves rapidly in positive feedback to approach a balance. Soon stability is achieved and the planet runs on in comfortable homeostasis.

The atmosphere of this new model of the Archean would be like a somewhat diluted version of the gas above a septic tank or a biogas generator—smelly and toxic for us, but delightful for the denizens of those ancient times. The atmospheric abundance of both carbon dioxide and methane would range between 1.0 and 0.1 percent. It is interesting that H. D. Holland was doubtful about the continuation of an atmosphere with a high content of carbon dioxide for long into the Archean. The rates of rock weathering from the geological record are not consistent with the persistence of 10 percent or more carbon dioxide. The rapid removal of carbon dioxide by the bacterial ecosystems neatly removes this problem. It is worth noting that many kinds of bacteria, not just photosynthesizers, actively remove and use carbon dioxide and make chemical compounds from it.

According to the model, the atmosphere was wholly different in composition in the Archean after life began. Table 4.1 illustrates the mixing ratio of the principal atmospheric gases before and after life. It shows an increase in nitrogen abundance after life began: I speculated that, until then, some of the nitrogen

Table 4.1 ESTIMATE OF THE ARCHEAN ATMOSPHERIC COMPOSITION BEFORE AND AFTER LIFE APPEARED

GAS	BEFORE LIFE	AFTER LIFE
Carbon dioxide	dominant	0.3%
Nitrogen	unknown	99%
Oxygen	0	1 ppm
Methane	0	100 ppm
Hydrogen	some	1 ppm

was present as the ammonium ion $(NH_4)^+$ in the oceans. The sea was more acid from the excess of carbon dioxide, and was rich in ferrous iron. In these circumstances, the ferrous iron may well have sequestered a large proportion of the ammonium ion to make a stable iron-ammonia complex compound, in which form much of the element nitrogen would have existed. Both the fall in carbon dioxide and the use of nitrogen by life could have changed the balance in favor of nitrogen gas in the air. Although nitrogen has no greenhouse effect by itself, the increase in nitrogen would have doubled the atmospheric pressure and this would have increased the greenhouse effect of the carbon dioxide and methane gases. The reason for this is somewhat recondite, but is connected with an increase in the amount of infrared absorbed by the greenhouse gases when total atmospheric pressures are higher.

It is important to note that there are other equally plausible models of the Archean. The conventional wisdom is expressed in Holland's book; it sees the pre-life environment continuing unchanged. Robert Garrels prefers to see the period as one where there were high temperatures sustained by high concentrations of carbon dioxide in the air. It is likely to be a long time before we are certain about the ancient history of the Earth. The purpose of this chapter, however, is not to make a firm statement on conditions during the Archean; it is to show how Gaia theory can be used to build from the meager evidence a different picture of those times.

I like to imagine some alien chemist arriving in the Solar System long ago and viewing the Earth's pre-life atmosphere.

The infrared spectrometer aboard the spacecraft would recognize a planet in the abiological steady state—a planet not yet alive, but with the potential to bear life. On a second visit much later in the Archean, when life had taken charge, a similar analysis would show a degree of chemical disequilibrium impossible for a lifeless planet. Carbon dioxide, methane, hydrogen sulfide, and oxygen cannot coexist at the levels shown in table 4.1 in the presence of sunlight. Given the destructive effects of solar ultraviolet radiation on methane, oxygen, and hydrogen sulfide, the alien would know that there was a large, continuous source of these gases. No conceivable volcanic source could sustain such an atmosphere. The alien would conclude that the Earth was now alive.

I often wonder what the Archean Earth would have looked like to us. I suspect that from an orbiting spacecraft we would not have seen the familiar blue-and-white sphere with glimpses of land and sea beneath the aerial canopy. More likely, the view would have been of a brownish-red, hazy planet; like Venus or Titan, too obscured to see the surface below. The sky that now we see as blue and clear results from an abundance of oxygen. Oxygen is the permanent bleach that clears and freshens the air.

On a beach on the edge of an Archean continent, we would see waves breaking on smooth sand, and sloping dunes behind. It would be familiar except for the colors. The Sun high above would have an orange glow more like sunset. The sky would be a pinkish hue, and the sea, that great copyist, shades of brown. There would be neither shells nor tracks of moving things upon the sand. The breakers offshore would fall away at low tide, exposing reefs of the strange mushroom-shaped stromatolites formed by the calcium carbonate secreted by colonies of living cyanobacteria. Inland, behind the sand and shingle dunes, would be flat and stagnant water, with patches of matted green and black bacterial growth. Other than the wind and waves, the only sound would be the plop of methane bubbles bursting as they broke from containment in the mud. Beyond the lagoon and on the continental surface, the same scene would repeat wherever there were shallow depressions in which water could gather. On the drier land and on the hill sides, a thin varnish of microbial life would ceaselessly work at weathering the rocks, releasing nutrients and minerals into the flow of rain water, and continuously removing carbon dioxide from the air. This quiet landscape could have persisted throughout much of the Archean. But there would have been violent interruptions when planetesimals crashed in from space. There were at least ten of these collisions; each a catastrophe great enough to destroy more than half of all planetary life. They would have changed the physical and chemical environment enough to hazard the remainder of life for hundreds if not thousands of years to follow. It is a tribute to the strength of Gaia that our planetary home was restored so promptly and effectively after these events.

Without life, the scene would have been much different. The ineluctable forces of chemical and physical evolution drive the small inner planets to an oxidized state through the loss of hydrogen. Venus must have had some water in the beginning. Estimates from the abundance of the unreactive noble gases suggests that, when the planets formed, Venus may have had at least a third as much water as the Earth. Where did it go? It seems most probable that the reducing elements of iron and sulfur in the surface rocks sequestered the oxygen of the water molecules. These reactions set hydrogen free as a gas, the light atoms escaping into space. The solar ultraviolet at the edge of the atmosphere may also have split some water vapor into hydrogen and oxygen. Either way, hydrogen, and hence water, was lost forever and the planet made more oxidized. Venus now, with its furnace heat and brimstone-laden air, is a model for Hell. By comparison, the Earth is Heaven for the life it bears.

How have we kept our oceans? It seems likely that the presence of life has done it. Robert Garrels tells me that his calculations suggest that, but for life, the Earth could have dried out in about 1.5 eons, midway through the Archean. There are several ways of retaining hydrogen on a planet. One is to add oxygen to the atmosphere or environment so that it captures hydrogen to form water. Life, in the act of photosynthesis, splits carbon dioxide into carbon and oxygen. If some of the carbon is buried in the crustal rocks, there remains a net increment of oxygen. For every atom of carbon buried, two atoms of oxygen are left behind. Each atom of carbon buried, therefore, is in effect four atoms of hydrogen or two molecules of water saved. Then there are the reactions at the ocean floor between sea water and the ferrous iron in basalt rock. The free hydrogen that these produce would be food for the bacterial species who could gain energy by using it to make methane, hydrogen sulfide, and other compounds less volatile than hydrogen. Methane, decomposed in the atmosphere by ultraviolet, could stratify the atmosphere and slow the rate of mixing of gases from the lower atmosphere, which would also hinder the escape of hydrogen to space. In these and other, more subtle ways, the presence of life in the Archean saved our planet from a dusty death.

Elso Barghoorn and Stanley Tyler first discovered the fossil bacteria that led to the recognition of the presence and the form of life in Archean times. I once visited Barghoorn's laboratory at Harvard University, and saw for myself the exquisite technical skills he used to cut, with diamond saws, the thin transparent slices of flinty rock. In this way, he and Tyler found the microfossils of bacteria in the ancient Gunflint rocks of the Great Lakes region of North America. But all these ancient fossils are from wet places, and we still do not know if there was life on the dry land. I find it hard to believe that a life form as enterprising as bacteria would have left unused the land surfaces. At this point, I should like to tidy away what I believe to be a persistent false assumption about those early times. We are using a new theory to view the scene; it helps to have the few genuine pieces of evidence displayed on a clean sheet.

The false image, that lingers like a mirage, is the shibboleth, "Earth's fragile shield." In a way, the atmospheric scientists L. V. Berkner and L. C. Marshall started it. Some thirty years ago, they introduced their famous theory on the evolution of atmospheric oxygen. Crucial to this was the assumption that there was a flux of lethal ultraviolet radiation before oxygen was present in the air and that this prevented life from colonizing the land surfaces. Indeed, it was further held that life before oxygen must have been obliged to exist deep in the sea at levels where the ultraviolet could not penetrate. It was only after oxygen appeared in the air that ozone could form and act as a shield to prevent the ultraviolet from reaching the surface. Once this happened, the way was open for an abundance of life to colonize the land and for the growth of oxygen concentration by increased photosynthesis to its present level of 21 percent. Some details of their theory we now suspect are wrong, such as that oxygen was at times more abundant than it is now. But this is no discredit; the information needed to test their theories was not then available. We owe an immense debt to Berkner and Marshall for the stimulating effect their ideas had on the development of the Earth sciences. Like Vernadsky and Hutchinson before them, they were scientists who presented a world model in which life had a part to play and was not just a spectator obliged to adapt to the climatic and chemical whims of a purely physical and chemical world. The scientific establishment accepted their ideas enthusiastically. Among those ideas was the minor postulate that the presence of a stratospheric ozone layer is an essential requirement for surface life. Almost every scientist now accepts it as if it were a proven fact of science.

There could have been no ozone layer at the start of life and during the Archean; gases like hydrogen and methane were dominant in the atmospheric chemistry, and even if there had been some oxygen in the atmosphere it could not have been used to form ozone. (Ozone is produced when ultraviolet radiation in the stratosphere splits molecules of oxygen into two separate atoms, which then combine with other molecules of oxygen to form a three-atom variety of oxygen: O_3.) The intensity of ultraviolet in the absence of ozone would have been 30 times higher than is now incident upon the Earth's surface. Such an irradiation, it is said, would have sterilized the land surfaces. The more committed believers in the potency of ultraviolet hold that 10 to 30 meters of ocean water are needed to filter out the deadly radiation. Life, they say, could not have existed in shallower depths of the sea, let alone on the surface.

Much more probably, "Earth's fragile shield" is a myth. The ozone layer certainly exists today, but it is a flight of fancy to believe that its presence is essential for life. My first job as a graduate was at the National Institute for Medical Research in London. My boss was the courteous and distinguished generalist, Robert Bourdillon. I was privileged to watch, and later participate in, the experiments that he and my colleague, Owen Lidwell, made as they tried to kill bacteria by exposing them to unfiltered ultraviolet radiation. Our practical objective was the prevention of cross infection in hospital wards and operating theatres. We were seeking a way to kill airborne bacteria and so prevent the spread of infection. Naked washed bacteria of some species, when suspended in the air as fine droplets, were easily destroyed by ultraviolet. It was impressive, though, how small a film of organic matter would almost entirely protect even these sensitive species. In the real world outside the laboratory, bacteria do not exist suspended in distilled water or a saline solution. In their normal habitats, bacteria are clothed in mucus secretions or the organic and mineral constituents of their environment. They do not live naked anymore than we do. Many practical trials were made before it was realised that ultraviolet radiation is not an effective method of eliminating from the hospital environment the tender fragile pathogens. It takes almost no clothing to stop ultraviolet radiation.[*]

The memory of these experiments left me disinclined to accept that the much weaker irradiation of the land surfaces in the Archean by natural ultraviolet could have prevented their colonization. The organisms then around were used to living outdoors in the sunlight and had millions of years in which to adapt themselves or the Earth. It is also wrong to assume that ozone alone among atmospheric gases can filter out ultraviolet light.

[*] Those still skeptical might be persuaded by the reports of these experiments in the Medical Research Council's special report number 262, entitled "Studies in Air Hygiene" and published in 1948.

Many other compounds absorb and remove shortwave ultraviolet radiation. The most probable candidates in the Archean would be the smog-like products from the decomposition of methane or hydrogen sulfide. In the ocean there are even more possibilities. The abundant ions of such transition elements as iron, manganese, and cobalt are intense absorbers of ultraviolet, as are the ions of nitrous acid and of many organic acids. But even if the full unfiltered solar ultraviolet shone on the surface, it still would not have much hindered life. Organisms are nothing if not opportunistic. They would probably have turned the hard ultraviolet light to use as a premium energy source. It is an insult to the versatility of biological systems to assume that a weakly penetrating radiation like solar ultraviolet could be an insurmountable obstacle to surface life. Even dark-skinned humans are almost immune to its effects; and it is used in the skin of us all for the opportunistic photobiochemical production of vitamin D.

This belief that ultraviolet radiation is unconditionally lethal to life on Earth has sustained a distorted view of the Archean and of other periods in the evolution of Gaia. And it is a view still deeply entrenched in scientific thinking. I found it to be common among the scientists who sought life on Mars. I could not help wondering how they could think that there was life on the intensely irradiated surface of Mars and at the same time believe that the land beneath the thick and murky Archean atmosphere of Earth was sterile. How could they fit into their minds two such contrary ideas?

I think that a more serious threat to the health of land colonies in those times would be the need for rain. Rainfall on the continental land masses of the present Earth is, to a considerable extent, a consequence of evapotranspiration: the pumping by trees and large plants of water from the soil to their leaves where it evaporates. The rising plumes of water vapor over forests act like invisible mountains and force the inflowing air from the oceans to rain out its burden of water. Even if bacterial life grew to form stromatolites, it is unlikely that these colonial structures that rose above the surface would be as efficient at rain making as trees. (Small though they are, however, bacteria do have tactics for rain making. Recently, scientists have found that bacteria of Pseudomonad type synthesize a macromolecule which can induce freezing in water droplets supercooled below 0°C.)

Although bulk water, as in a swimming pool or even in a glass, freezes when its temperature falls below 0°C, droplets of water that have condensed inside a cloud may not freeze until the temperature falls to −40°C. This supercooling takes place in the absence of nuclei of solid particles onto which the first microscopic ice crystal can form and grow. Pure water is reluctant to freeze; it freezes in our refrigerators because, in bulk water, there is always at least one nucleus to start the process of nucleation. Some chemical substances, such as silver iodide, have crystals close enough in shape to mimic ice. If these are dusted on a supercooled cloud, they will start the freezing process and sometimes the fall of rain. The macromolecule that the pseudomonads synthesize can cause droplets cooled only to −2°C to freeze, and is far more efficient than silver iodide. (This has led to commercial interest in methods of rain making. Silver iodide crystals work after a fashion, and production of the efficient pseudomonads macromolecule is under way. But it is thought by some environmentalists to be socially undesirable; the stealing of rain that might otherwise have fallen on those who may have needed it more.)

Pseudomonads have an ancient history, and maybe their ice-nucleation trick goes back to the Archean. If so, were they the rain makers that led the colonization of the land? A question that always arises at this point in speculation is: How did it happen? Surely the bacteria did not decide to make the ice-nucleating substance. At this point, serious-minded microbiologists grow anxious and fear the proximate occasion of teleological heresy. Fortunately, we can easily make a plausible model of the evolution of close coupling between a large-scale environmental effect and the local activity of microorganisms—a model, moreover, free of any taint of purpose.

It is probable that the regional and global physiological systems of Gaia have their origins in some local competition and negotiation between species. An early variant of the ice maker may have found that the freezing of dew at its growth site gave some advantage. It might have been destroying a competitor or predator by freezing, splitting the tough skin of a food organism, or producing mechanical fractures in rocks to release nutrients or increase the quantity of soil particles. Any of these effects, alone or in combination, would confer advantage on the ice maker and, more important, favor those that made the most or the best nucleator. Eventually, the best possible nucleator would be ubiquitous in its distribution. For purely local reasons, these bacteria would continue their freezing activity wherever it was to their advantage. It is not difficult to see that surface ecosystems carrying ice makers would be at an advantage under drought conditions compared with those unable to produce the nucleating agent. The soil dust stirred by the wind or lifted by whirlwinds could induce droplet freezing in the clouds and then rainfall.

The connection between the freezing of cloud droplets and the subsequent fall of rain is well understood. A great amount of heat is released when water freezes; in other words, freezing half of the water in a drop supercooled to $-40°C$ releases enough heat to raise the temperature of the mixture of water and ice by $40°$ to the freezing point. If a large proportion of supercooled droplets in a cloud freeze, the latent heat released warms the cloud and causes it to rise. More water vapor condenses and freezes so that ice and snow falling though the cloud gather water and weight, and fall as rain. Any product of living organisms that nucleates supercooled cloud droplets will therefore encourage rain.

More important in climate regulation than the nucleation of supercooled water droplets is the nucleation of supersaturated water vapor. The air above the open ocean is often supersaturated with water vapor. But no clouds or moist droplets can form until fine particles, the cloud condensation nuclei, appear. The climatologist Robert Charlson has argued that the emissions of sulfur compounds by the biota now and in the recent past has played an important role in providing cloud condensation nuclei.

But this requires the presence of atmospheric oxygen to oxidize the sulfur to sulfuric and methanesulfonic acids, the nucleating agents. This could not have happened in the Archean, but there may have been other molecular species that served in this way. The aerosol of sea salt from breaking waves has some capacity to nucleate clouds, but it is slight compared with that of the sulfur acid micro-droplets.

Although rainfall is essential for growth on the land, it also poses problems because it washes away nutrients. (The poor productivity of the rain-washed uplands of the west coast of the British Isles is an example of this problem.) Today, rivers carry to the ocean elements that are used or required by marine life—such as nitrogen, phosphorus, calcium, and silicon. But the rivers also carry the rarer elements—sulfur, selenium, and iodine—to the sea, and the land becomes depleted. This brings us to another large-scale geophysiological mechanism: the transfer of essential or nutritious elements from the ocean, where they are abundant, to the land, where they are scarce. The process requires marine life to synthesize specific chemical compounds that act as carriers of the elements through the air. The element sulfur, for example, is carried from the ocean to the land by dimethyl sulfide, a product of marine algae. In the Archean, the environment was either oxygen-free or there was an excess of reducing gases over oxygen. In such an atmosphere, the synthesis of dimethyl sulfide, which seems to take place only in oxic environments, is unlikely. Compounds such as hydrogen sulfide and carbon disulfide, which are unstable in our present oxidizing air, could have served instead to carry the essential element sulfur, also in the Archean land life could have needed less.

Hydrogen sulfide is ubiquitous in the anoxic zones and reacts with many metals—such as lead, silver, and mercury—that might otherwise accumulate to toxic levels. The result is water-insoluble sulfides that settle as solids. The geochemist Wolfgang Krumbein has shown that the ore beds of these elements exposed on or near the surface today are the waste tips of some past anoxic ecosystem. Anaerobic organisms that converted the potentially toxic elements, mercury and lead, to their volatile methyl derivatives grew successfully and provided the ecosystem with a mechanism to remove toxic waste. The anoxic zones are continuously perfused by a flow of methane gas that would serve to carry these volatile materials away from the region. Some of this methylating activity is beneficial on a regional or even global scale. The production of dimethyl selenium serves in a subtle way, first discovered by the atmospheric chemist F. S. Rowland, to offset the toxicity of dimethyl mercury. It also acts to recirculate the essential element selenium through the global environment.

The rate of carbon burial during the Archean was not significantly different from today. As we saw earlier, the carbon present in the earliest sedimentary rocks shows a subtle difference in the proportion of its isotopes from that of lunar rocks that have never been exposed to life; this difference is evidence for the presence of photosynthesizers. The geologist Euan Nisbet tells me that there are Archean, carbon-rich shale deposits in southern Africa. They are like the coal measures put down by the forest trees of the Carboniferous period, eons later. These carbon deposits are all that remains of the dead bodies of microorganisms that once grew in the Archean. Volcanoes then, as now, vented carbon dioxide. Archean photosynthetic and other bacteria used this carbon dioxide to make the organic compounds of their cells; these organisms also may have facilitated the reaction of carbon dioxide with calcium and other divalent ions dissolved in the sea and on the land surfaces. These two reactions were the sinks for carbon dioxide and kept a steady level in the atmosphere. This is part of the climate regulating system illustrated in figure 4.2. In addition to these climatic consequences, the Archean ecosystems would have buried a small but constant proportion of their carbon turnover, which would have led to the steady addition of oxygen. This, however, would have been used up in oxidizing the reducing compounds of the surface and ocean environments, and that emitted by volcanoes. It was somewhat like one of those chemistry experiments in high school, where you progressively add an oxidizing solution to a reducing solution until an indicator suddenly changes color to mark the equally sudden change from reducing to oxidizing at the end of the titration. The burial of a small proportion of the carbon and sulfur, cycled by once-living bacteria, titrated the oxidizable material of the environment until the surplus was used up. Reducing material continued to be added to the ocean and the atmosphere, but the rate of its addition became less than that of carbon burial. Free oxygen gas began to appear in the air at levels more than sufficient to overcome the reducing tendency of methane, and marked the end of the epoch.

It seems likely that the end of the period when methane dominated the chemistry of the atmosphere was abrupt. But it would be wrong to envisage a sudden change from a wholly oxygen-free world to one where oxygen was present free in the air. Much more probable is the gradual growth during the latter part of the Archean of oxic organisms at the surface of the Earth. These could have existed first at the surface, where the phototrophs basked in the sunlight and locally produced enough oxygen to support them. They would be a separate and encapsulated ecosystem surviving in an otherwise lethal system, rather as the anaerobes survive in the poisonous oxygen-rich world of today. In this oxic ecosystem there would be consumers living on organic products of the cyanobacteria, and also organisms able to exploit a slightly oxidizing medium and perform such tricks as denitrification (using nitrate and nitrite ions instead of oxygen, so that nitrogen escaped to the air as gaseous nitrogen and nitrous oxide).

Gradually, as the oxygen-scavenging compounds of the sea were used up, the oxygen released by the phototrophs would no longer be absorbed. Then the ratio of the methane to oxygen flux to the atmosphere would shift towards an oxygen excess. The oxic ecosystems would spread and, just before free oxygen increased in abundance to become the dominant oxidizer, would probably have covered most of the oceans. The changeover was not so much a genocide as a domination. Even stranger scenarios are likely if the surface communities generated nitrous oxide before oxygen itself appeared. This gas is stable in the troposphere and might have allowed methane to persist longer; it is also somewhat of a greenhouse gas and might have compensated for the decline in methane. It is made by bacteria now, and it is likely that there were bacteria making it then.

In geophysiology, the Archean boundary coincided with the great punctuation marked by oxygen's free presence in the air. However, for the bacteria of the Archean the era never ended. They live on wherever the environment is free of oxygen. They run the vital and extensive ecosystems of the anoxic zones beneath the sea floor, in the wetlands and marshes, and in the guts of nearly all consumers including ourselves. In a strict geological sense, the period ended 2.5 eons ago, and oxygen may have come later. The appearance of oxygen in the air and on the surface of the oceans did not eliminate the anoxic ecosystems; it merely segregated them in the stagnant waters and sediments. As a consequence, the rocks that formed from these sediments may have failed to record the presence of free oxygen in the air.

That, then, is an account of a few aspects of the Archean seen through Gaia theory. It was a period when the Earth's operating system was populated wholly by bacteria. It was a long period, when the living constituents of Gaia could be truly considered as a single tissue. Bacteria are both mobile and motile, and could have moved around the world carried by winds and ocean currents. They can also readily exchange information, as messages encoded on low-molecular-weight chains of nucleic acids called plasmids. All life on Earth was then linked by a slow but precise communication network. Marshall McLuhan's vision of the "global village," with humans tied in a chattering network of telecommunication, is a re-enactment of this Archean device.

5

The Middle
Ages

*Little is known about the important transitional period from
about 2.0 billion years ago until about 0.7 billion years ago.*
ROBERT GARRELS

If you are curious about the Earth and wonder about the history
of rocks, there are few better places to be born than England
or Wales. My small island, which lies within both of these
countries, has as many geological periods as a continent. Along
its lengthy shoreline the waves have cut cliffs and these walls
of rock display their dissected strata as in a museum diorama.
I used to spend childhood holidays at a place called Chapman's
Pool on the coast of the county of Dorset; here the somber
black cliffs of Kimmeridge shale were speckled with snow-white
ammonites and other fossils.

As you move westward across England the rocks go back
in history, and by the time you come to Wales their age approaches 570 million years. These old rocks are called Cambrian,
after the Roman name for Wales. They are the oldest to bear
fossils visible to the unaided eye. There are, of course, older
rocks containing microfossils of bacteria, such as those that

Barghoorn and Tyler found; but before modern methods of dating there was no sure way of knowing their age. The period with rocks older than those bearing the larger fossils was called Precambrian, because it was further distant in time than the rocks of Cambria. We now know that the Precambrian has parts that are very old indeed. This new knowledge comes from the distribution in these old rocks of the radioactive elements uranium and potassium, and their products, lead and argon. Radioactive decay is an accurate clock; by measuring the proportion of uranium to lead or of potassium to argon in a piece of rock its age can be calculated. Other evidence about ancient rocks comes from the distribution of the isotopes of that stable element, carbon. This, and the discovery of Archean bacterial microfossils, tell us that life was present at least 3.6 eons ago. The Precambrian is now mapped and divided into the Proterozoic period, 0.57 to 2.5 eons before now, and the Archean, 2.5 to about 4.5 eons. Some geologists call the first period, 4.5 to 3.8 eons, the Hadean.

Like the Archean, the Proterozoic was a time when the ecosystems of the Earth were populated by bacteria (the prokaryotes). In the anoxic regions of the sediments the Archean bacteria would have lived on; but in the now mildly oxidizing ocean and surface environments there eventually developed more complicated living cells, the eukaryotes. These are the ancestors of large communities of nucleated cells, like the trees and ourselves.

The Proterozoic is still an enigmatic period of the Earth's history. I feel free, therefore, to use it as a background on which to develop geophysiological models of what it might have been. I shall write this chapter not as history but as an account of the physiology of an unknown animal who lived thousands of years ago, drawn from no more evidence than a few scraps of bone accurately dated from their content of carbon isotopes. My main interest is in the long-term geophysiological processes that kept the Earth constant and fit for life, and in how they worked. With the bones of the Earth in mind, I shall keep returning in thought to that important element, calcium, and its crucial role in all living things from ourselves to Gaia. I shall continue to be concerned with those other important elements, oxygen, carbon, and hydrogen, and with their regulation and with the climate. This chapter will also be about the geophysiology of the oceans; in particular the difficult problem of whether the total salinity is determined by physical and chemical forces alone, or whether there is "machination" on the part of Gaia. Although the setting of the chapter is the Proterozoic, the middle ages of the Earth, most of the topics discussed are not unique to that period; they acted also in the Archean, and continue to act in the present period.

If we aim to start at the boundary between the Archean and the Proterozoic, we shall find that this boundary is still under negotiation. There is no clear-cut frontier, just a no man's land where field geologists set their posts according to their fancy. The Archean geologist Euan Nisbet tells me that there is an informal acceptance of the date 2.5 eons before the present time, although some would prefer to set the fence at the date marked by the appearance of a special suite of rock in Zimbabwe. Just as political frontiers often fail to circumscribe ethnic regions accurately, so boundaries based on geological considerations alone do not always suit the interests of geophysiology. As a geophysiologist, I prefer to set my markers at the time when the environment became predominantly oxidizing—or to put it more professionally, at the transition from an environment dominated by electron donor molecules, like methane, to one dominated by electron acceptors, like oxygen. As it happens, the uncertainties about the events 2.5 eons ago are large, and for the time being we can take the geological and the geophysical markers to define the same zone of the Earth's history.

For geophysiology, the important thing about the transition from the Archean to the Proterozoic is not the exact date of the event, but that it happened. It is rather like puberty; a profound physiological change but one spread over a finite period of time. In puberty markers of the change—the appearance of the beard and the deepening of the voice, or the expansion of the breasts—are secondary to the main event. The switching on of these secondary sex characteristics is the response to an increasing flux of pituitary hormone. This primary event may be sharply defined but the secondary characteristics are somewhat arbitrarily scattered in time. Between the Archean and Proterozoic the appearance of oxygen as a dominant atmospheric gas was the primary event and marked a profound change in the Earth's geophysiological state. The outward and secondary manifestations of this change—the emergence of a new surface and atmospheric chemistry and of ecosystems as oxygen began to dominate the atmosphere—is likely to have spanned a significant interval of time, and happened at different times in different places.

From early in the Proterozoic to the present day there has been an excess of free oxygen gas in the air. By an excess, I mean that the atmosphere has carried more oxygen than is needed to oxidize completely the short-lived reducing gases: methane, hydrogen, and ammonia. When the Proterozoic began, geophysiologically speaking, the division of the two great planetary ecosystems, the oxic surface regions and the anoxic sediments, was complete. The Archean, when the environment was full of molecules that donated electrons (that is, reducing agents), did not so much end as become encapsulated as a separate region that exists wherever oxygen is absent. The submission of the anoxic ecosystems to domination by the oxic was somewhat like the Norman conquest with the Archean Saxons driven to a subservient underground position—the lower classes—from which, it is often said, they have never escaped.

The change from anoxic to oxic was a crucial step in the Earth's history. In the model of the Archean in the previous chapter (figure 4.2), the end of the period was pictured as very sudden, with oxygen rising from very low to between 0.1 and 1 percent atmospheric abundance in not more than one million years. This is, of course, no more than the prediction of a geophysiological model; one that sees the change from one regime to another as an event driven by powerful positive feedback from the biota and the environment.

The conspicuous difference between the Archean and the Proterozoic is, I think, in the composition of the atmosphere and the oceans; and possibly also in the climate. The simple model in figure 4.2 supposed that the carbon cycle was preserved by the methanogens which returned a massive flux of methane and carbon dioxide to the air from the sediments, and that this state persisted until, quite suddenly, free oxygen appeared. More probably there was some oxygen present even early in the Archean, just as there is methane in our present atmosphere. The difference between the Archean air and the Proterozoic air was not a simple matter of the presence or absence of oxygen, it was in the net tendency. In the Proterozoic, a discarded bicycle left in shallow water would have rusted away to form insoluble ferric oxide which settled on the sea floor; in the Archean, it would have slowly dissolved as water-soluble ferrous iron, and left no trace. During the time that the flux of methane exceeded the flux of oxygen, the lower atmosphere could carry only trace quantities of oxygen. The oceans and surface rocks, rich in the oxygen-scavenging ferrous iron and sulfides, would have absorbed so large a proportion of the oxygen output of the cyanobacteria that the air would have remained in a net anoxic state for most of the Archean period.

We do not know that the oxygen of the air rose suddenly; it may have risen slowly or in a series of steps. It is important also to distinguish between the presence, and the dominance, of oxygen; dominance, in a chemical sense, requires that the ratio of oxygen to methane is greater than two to one. The reason for thinking that the change of regimes, the boundary of the Archean and Proterozoic periods, was abrupt is the evidence of a major glaciation at about 2.3 eons ago. This might have come as a result of a sudden fall in atmospheric methane. Such an event would be accompanied by cooling, since methane and its decomposition products are greenhouse gases. There are also geophysiological arguments that favor a sharply defined transition to an oxidizing state. Once the photosynthetic oxygen became dominant in the atmosphere and oceans, the action of sunlight on oxygen would produce hydroxyl radicals that oxidize the methane in the air. Also, there would be consumers feeding on the organic matter before it could reach the anoxic sediments; this would deny to the methanogens the material for the production of their gaseous excretion. This is the recipe for a powerful positive feedback against methane and in favor of oxygen. These events are likely to have been sudden rather than gradual. Finally, there is the ecology to take into account. There would be ecosystems existing in the Archean that found their world comfortable. As they evolved with their environment, they would resist change, but their resistance would be like that of a locked fault in an earthquake zone. They would tend to resist change, trying to keep the status quo, but when the break came it would be all the more sudden and devastating.

The transition to oxygen dominance may have left its mark in the record of the rocks in the form of the Gowganda glaciation, but the long subsequent period is one of the more obscure periods of the Earth's history. Gaia theory requires that the tightly coupled evolution of living organisms and their material environment will have determined the state of the Earth in this as in the other periods. Can we envisage from this theory a living planet existing in the Proterozoic, and the regulatory systems that may have operated then?

When Earth scientists use the word regulation, they usually have in mind a passive process where the input and output of some component or property are in balance. In geophysiology, by contrast, regulation implies the active process of homeostasis; the preservation of a comfortable Earth by the interaction of life and its environment. The speculations that follow about the regulation of climate, oxygen, salinity and other properties of the environment are in this geophysiological context, in other words, as if they were speculations about the state of a living organism. In no sense is this intended as a teleology, or meant to imply that the biota use foresight or planning in the regulation of the Earth. What we need to think about is how a global regulatory system can develop from the local activity of organisms. It is by no means far-fetched to imagine a single new bacterium evolving with its environment to form a system that can change the Earth. Indeed the first cyanobacterium, progenitor of the ecosystem that used light energy to make organic matter and oxygen, did just this.

If the element oxygen was crucial in the geophysiological evolution of the atmosphere, then calcium must surely be the determining element in the geophysiology of the oceans and the crust. Calcium is one of the *alkaline-earth* elements occupying the second column of Mendeleev's famous periodic table of the elements. It comes after magnesium and before strontium. It is the third most abundant positive ion of sea water, after sodium and magnesium. We tend to think of calcium as a benign and nutritious element because it is an essential structural component of our bones and teeth. It is also crucial in numerous internal physiological processes from blood clotting to the division of cells. It is essential for life but, paradoxically, it is very toxic in the free ionic state. Within our cells a concentration of calcium ions exceeding a few parts per million is lethal, comparable in toxicity to cyanide, yet calcium ions are free in the oceans at a level ten thousand times greater.

In chapters 2 and 3, I explained the operation of the Daisy-world model in terms of the competitive growth of organisms when one environmental property is sharply circumscribed. Too much and too little are both uncomfortable; there is a preferred state between torrid heat and freezing cold, between excess and starvation. This is particularly true of calcium. Imagine some bacterium in the early oceans able to convert the abundant water-soluble calcium ions of its internal environment into insoluble calcium carbonate. This simple reaction would have effectively reduced the concentration of the potentially toxic calcium ions within the cell, by locking up calcium in a safe insoluble form. Such an action, if calcium were in excess as it usually is in the oceans, would have increased the organism's chances of survival and those of its progeny. These organisms would be at an advantage compared with organisms that merely adapted to the presence of excess calcium. In the sunlit zone of the open ocean, the growth of these organisms would lead to vast masses of calcium carbonate being deposited on the ocean floor. The rain of microscopic "sea shells," called tests by marine biologists, from the sunlit surface to the depths acts like a conveyor belt. Food is brought for consumers lower down, the ocean is swept clear and made transparent, and potentially toxic elements, like cadmium, are taken from the surface regions. Carbon dioxide and calcium are transported, and assembled together by bacterial communities to form the flat or mushroom-shaped rock cities called stromatolites. The concentration of calcium ions in the oceans would have been reduced, and all life would have flourished as a consequence. The ubiquity of limestone deposits of oceanic origin implies the success and continuation of this activity. In contradiction to this view, some geologists believe that early limestone deposition was an inorganic process. I do not see how we can distinguish between the spontaneous crystallization of super-saturated calcium carbonate in the Archean and nucleation induced by organisms. I do think that the nucleation of supersaturated and other metastable states in Nature is a key geophysiological process, and that it originated in the Archean.

I know as an inventor that really good inventions tend to grow and evolve. Only weak inventions take a single step and no more. Consider how the simple semiconducting crystal of those first radio receivers in the 1920s has evolved to become, in a grand eutrophication, the ubiquitous silicon devices of today. The calcium carbonate precipitation step was an even greater invention; it led not merely to the regulation of calcium, carbon dioxide, and climate but also to the vast engineering of the calcium carbonate structures (the stromatolites). Later, these same processes evolved so that our own cells possess intricate mechanisms by which calcium is deposited as bones and teeth.

Most remarkable of all, biological calcium carbonate deposition may have made possible the efficient operation of the *endogenic* cycle—the slow movement of the elements from the surface and the ocean to the crustal rocks and back again. The geologist Don Anderson has speculated that the deposition of limestone on the ocean floor is a key factor in the motion of the Earth's crust. He proposed that sometime far back in the Earth's history, sufficient limestone was deposited to alter the chemical composition of the crustal rocks of the ocean floor near the continental margins. As a result an event, called the basalt-eclogite phase transition by geologists, took place. This transition so altered the physical properties of the crustal rocks that it became possible for the great machinery of plate movement to begin turning. Don Anderson commented in an article in *Science* in 1984:

> The Earth is apparently also exceptional in having active plate tectonics. If the carbon dioxide in the atmosphere of Venus could turn into limestone, the surface temperature and those of the upper mantle would drop. The basalt-eclogite phase change would migrate to shallow depths, causing the lower part of the crust to become unstable. Thus there is the interesting possibility that plate tectonics may exist on the Earth because limestone-generating life evolved here.

To me this is an exciting idea, but I admit that most geologists find it extremely improbable. The event may have begun with the activity of a few organisms able to split a dilute solution of calcium bicarbonate into chalk and carbon dioxide, and so avoid calcium poisoning. We do not know when plate tectonics started. If it is connected with life, that connection may not have existed before the development of the intracellular precipitation of calcium carbonate in eukaryotes, late in the Proterozoic.

The study of the intricate biological processes for segregating and concentrating the elements of the crust and ocean in the form of minerals has become a separate topic of the Earth sciences called *biomineralization*.

Salt regulation is one of the most interesting and tantalizing Gaian systems. There are few organisms able to tolerate salt at concentrations above about 6 percent by weight. Have the oceans always kept below this critical limit of salinity by chance? Or has the tightly coupled evolution of life and the environment led to the automatic regulation of ocean salinity? It is often stated that the preferred internal saline medium of living things, one that is astonishingly similar over a very wide range of organisms, reflects the composition of the oceans when life started. It is true that the salinity of the blood of whales, humans, mice, and of most fish, whether dwelling in the ocean or in fresh water, is the same. Even the circulating fluid of *Artemia,* the brine shrimp that lives in saturated salt solutions, shares with us the same internal salinity. But to my mind this is no more evidence of the salinity of the Archean ocean than are the oxygen levels now breathed by these organisms evidence of the oxygen abundance at the start of life.

Most cells survive and do best in a medium whose salinity is 0.16 molar (about one percent by weight in water, or *normal saline*). Many kinds of cells survive the salinity of sea water, 0.6 molar, but above 0.8 molar the membranes that hold the precious interior contents of cells become permeable or disintegrate completely. The reason for the destructive action of salt solutions is simple. Cell membranes are held together by the same kind of forces that hold together a soap bubble. Quite often these forces are very sensitive to the salinity of the medium, usually being weakened when the salinity is high. You can see this for yourself by making bubbles with increasingly strong salt solution as the solvent for the soap. Above about 10 percent salt, bubbles cannot be made. This is because soap molecules are made up of a long chain of carbon atoms tightly linked together and surrounded by hydrogen atoms equally firmly held. These chains are terminated at one end only by a carbon atom with two oxygen atoms attached. When this end of the soap molecule is dissolved in water that is slightly alkaline, as all soap solutions are, it carries a negative charge. The negative electrical charge makes the end group attract water molecules and drags the insoluble oily hydrocarbon chain into solution. Salt is made up from negative chlorine ions and positive sodium ions. When there are many of these in solution they begin to compete with the soap's negative charge for the water molecules. When enough salt is present, the soap separates from the water as curd.

The molecules of a cell membrane are more complex than

soap: they include such substances as sterols (for example, choles-terol), hydrocarbons, proteins, and phosphatides. The molecules of phosphatides correspond to those of soap, and are the part of the membrane most affected by an increase in salt concentra-tion. High salt concentrations disrupt cell membranes by disturb-ing the electrical forces holding the membrane in its correct and complex state. The membrane of the human red blood cell, for example, is made up of more or less equal parts of cholesterol and lecithin (a phosphatide) and a mixture of protein and other fatty substances. The red blood cell will survive exposure to salt solutions up to 0.8 molar (4.7 percent by weight in water). Above this strength the membrane is damaged, and at concentra-tions above 2.0 molar the cell may be destroyed in seconds. In the mid-1950s I was able to show by direct experiments that the damage to red blood cells commenced when lecithin dissolved away from the membrane into the strong salt solution. This kind of salt damage seems general among living cells and is seen with the cells of all five kingdoms.

The salt concentration of today's sea is always uncomfortably high for living organisms. The larger ones, such as fish, swimming mammals, and some crustacea, have physiological mechanisms to regulate the internal salt at a level close to that of their own bodies (0.16 molar). To prevent the loss of water from their internal medium (osmosis), these animals have to work to stop themselves being squeezed dry by the osmotic pressure. They have to use energy to pump water in against the osmotic pressure difference between their interior and that of the sea. The pressure is close to that needed to pump water 450 feet vertically against gravity, over a thousand times the blood pres-sure of humans. The advantages of a low-salt interior must be considerable to require such an effort to sustain it. For the small cell of a bacterium, individual regulation is a luxury far beyond its means. This does not just apply to salinity. Take temperature for example: for a microorganism to sustain even a 1°C difference from that of its environment would require the consumption of far more food and oxygen than could cross its surface. Not only this, but a 1° temperature gradient across the cell membrane would generate a thermal osmotic pressure of 56 atmospheres, or 840 pounds per square inch; far beyond the strength of a cell membrane to resist.

The most frequent experience of salt stress is in drying or freezing. When cells freeze, water is removed as pure ice and the salt solution in which they exist becomes concentrated. Freezing and drying must have been common hazards from the beginning of life, and indeed no direct answer to this problem of salt damage to membranes has evolved in all the time since then. There are salt-tolerant bacteria, the halophiles, that live precariously in the saline regions of the Earth. These bacteria have solved the problem directly by evolving a special membrane structure that is not disrupted by salt. It works, but at a price; for these organisms cannot compete with mainstream bacteria when the salinity is normal. They are limited to their remote and rare niche, and depend upon the rest of life to keep the Earth comfortable for them. They are like those eccentrics of our own society whose survival depends upon the sustenance that we can spare but who could barely survive alone.

Mainstream life, therefore, is limited to a maximum salt concentration of about 0.8 molar. This is not much greater than the saltier parts of the ocean, where the salinity reaches 0.68 molar. It is frequently exceeded whenever the tide recedes, leaving organisms soaked in sea water to dry out on the shore. The problem of salinity must have arisen for the Archean bacteria. Their answer to the problem was to synthesize soluble compounds called the sulfur and nitrogen betaines. They make these neutral solutes, which substitute for the salt and are not toxic to the cell. When these are present in the cells and their medium, freezing or drying no longer concentrate the salt to destructive levels. Even so, there is a price to pay for the synthesis of these anti-salt betaines. As much as 15 percent of the dry weight of shoreline algae is betaine; a considerable diversion of the energy that would otherwise be available to the organisms. Clearly, it would be to the advantage of the biota to keep the oceans as dilute as possible—certainly to keep them from approaching the critical 0.8 molar concentration.

Problems of salt regulation exceed in difficulty anything humankind has so far done in the way of planetary engineering. The Archean bacteria could relatively easily modify that small compartment, the atmosphere, to suit their needs; but the vast mass of the oceans, some ten thousand times larger, was very much more difficult to manipulate. The only way to remove the huge masses of salt in the oceans would be to segregate ocean water in lagoons and allow the sun's heat to evaporate the water away. It would have required the building of vast limestone reefs to trap salt in evaporite lagoons. The sheer magnitude of these reefs would dwarf any conceivable human construction. It may even be possible that the process of lagoon formation was assisted by the folding of rocks at the continental edge as a consequence of the movement of the plates.

Table 5.1 lists the salts of the oceans. Salt in solution, weathered from the rocks, is entering continuously from the rivers and also from the interior of the Earth at the sea-floor spreading zones that lie at the bottom of all the major oceans. In the ocean, salt is not a single substance, sodium chloride. Rather, it exists as positive sodium and negative chlorine ions, and these behave as two quite independent and separate entities. The sodium ion and the other positive ions of potassium, magnesium, and calcium all have relatively short residence times in the ocean. They are removed by biochemical and chemical processes, and also by hydrothermal chemical reactions within the sea-floor spreading zones, and deposited as sediments, clays, lime-

Table 5.1 SALTS OF THE OCEANS

IONIC SPECIES	ABUNDANCE (MOLES)	RESIDENCE TIME (MYR)
Sodium	0.47	56.0
Magnesium	0.05	11.0
Calcium	0.01	0.9
Potassium	0.01	5.5
Chloride	0.53	350.0
Sulfate	0.03	7.9

SOURCE: M. Whitfield, 1981, "The World Ocean: Mechanism, or Machination," *Interdisciplinary Science Reviews* 6, 12–35.

stones, and dolomite. The salt problem for the biota is actually a problem of getting rid of the negative chloride ions.

Chemically, the chloride ion is rather like an atom of the wholly unreactive gas argon. It is a smooth, slippery, spherical molecule with little or no tendency to attach to anything. There is no significant biochemical trade in chlorine. A few strange systems do make methyl chloride from salt, but the turnover of this compound is too small to affect the salinity of the ocean. Also, the chlorine in methyl chloride soon becomes chloride ions again, and is washed out by the rain and returns to the sea. Chloride ions are removed from the oceans physically by the transfer of sea water to the evaporite lagoons. The water trapped in these lagoons warms in the sun and evaporates; the water vapor moves through the atmosphere and eventually condenses elsewhere as rain—pure water—that flows into and dilutes the ocean. The salt, represented by the dominant chloride ion and the positive ions that must go with it to keep the total ionic charge zero, is left behind as crystalline layers. These evaporite lagoons are found in many places on the continental margins. Fossil lagoons exist in many places beneath the Earth's surface, sometimes even beneath the ocean itself.

Before these lagoons can form, barriers are needed at their seaward boundary. Could this activity be part of the tightly coupled evolution of life and the rocks, or is it just the result of chance? The key process in the formation of these barriers is the deposition of calcium carbonate. The carbon dioxide in the air reacts continuously with alkaline rocks on the land surfaces to form bicarbonates. An important reaction of this type is the one between calcium silicate rock and carbon dioxide dissolved in surface water. The product is a solution of silicic acid and calcium bicarbonate, which flows down the rivers to the ocean. In the absence of life, the calcium and bicarbonate ions can coexist in a mildly acid ocean, and the continuous supply of these would eventually lead to the spontaneous crystallization of calcium carbonate. But it would be more or less randomly deposited over the ocean floor. Limestone deposits in the real world are mostly from the action of living organisms.

Limestone is not deposited either randomly or according to the expectations of physics and chemistry. The precipitation of calcium carbonate by colonies of microorganisms occurs most extensively in the shallow waters around the continental margins where the abundance of both nutrients and calcium bicarbonate are highest. Without any planning or foresight, the components of those living structures, the limestone stromatolites, would have assembled offshore and eventually sealed off lagoons from which sea water would be progressively evaporated and salt deposited. At first the reef building would have only a local effect, but over time the sheer mass of the limestone would begin to affect the plastic crust of the Earth's surface, depressing it and so extending the size of the lagoon. New rock formers would always be colonizing the surface of a reef as it descended, so tending to keep the lagoon intact. If, as Don Anderson has suggested, the motion of the Earth's crust depends on the continued deposition of calcium carbonate in the sea, the limestone reefs could have led to the complex events of mountain building and the folding of rocks at the continental margins. This, in turn, would extend the range of shorelines where evaporite lagoons could form.

During the course of time salt is being added to the oceans from the lithosphere and removed again. Some of this salt is deposited in *evaporite beds* and buried beneath sediments. These deposits may be a temporary store which Earth movements and weathering continuously expose and release the contents of back to the sea; but new evaporite lagoons are always forming. The balance of erosion and formation seems always to have kept enough salt sequestered in evaporite beds to keep the oceans fresh and fit for life. The evidence that lagoon formation and maintenance depends on the specific behavior of marine microorganisms is strong.

I once had the pleasure of joining an expedition, led by Lynn Margulis, to the microbial mat communities that form in the evaporite lagoons of Baja California in Mexico (as shown in figure 5.1). They are on the western edge of that long, narrow tongue of land that hangs down below San Diego and separates

5.1 The evaporite lagoons at Laguna Figueroa, Baja California, Mexico. Sand dunes form a barrier through which sea water percolates. The water then evaporates in the lagoons, and salt crystallizes out to form an evaporite deposit. The surface of the lagoons are often covered by microbial cell communities called mats.

the Gulf of California from the Pacific Ocean. Here I was able to see at first hand the subtle economy of the bacterial mats that covered the lagoon. The red and green communities of microbes at the surface acted as a raincoat, preventing the salt from being dissolved in the rain and washing back into the ocean. Indeed, on one occasion, the whole lagoon was flooded by feet of fresh water. Within two years the flood was evaporated and dispersed without destroying either the microbial communities or the evaporite bed beneath. In normal times, the movement of rain water downward through the mat lowers its salinity and assists the growth of photosynthesizers at the surface, the primary providers of food and energy for the communities beneath. Salt crystals at, or near, the surface are also coated with their own specific varnish and protected against easy solution in rain water.

Is all this a grand, unplanned civil engineering enterprise by Gaia? The steps, from the individual lowering of calcium ions within the cells of a living organism to the movement of the plates, are all those that tend to improve the environment for the organisms responsible. But the links between biomineralization, salt stress, and plate tectonics are so tenuous that most scientists would think them to be connected by chance rather than by geophysiology. I shall continue to wonder about the limits of Gaian manipulation and always seek guidance by asking the simple question: What would the Earth have been like without life? Would limestones have precipitated at the continental margins so as to form evaporite lagoons? Would the salt have deposited in them, or would it have been washed back into the ocean by rain in the absence of a living raincoat of microbial mat? Would limestone have deposited at the sites and densities needed to start the plates moving? Unlikely perhaps, but remember there have been billions of years for geophysiological invention and its trial by natural selection. We should consider the possibility that this long period was sufficient to fine tune the rough geology into a smoothly regulated geophysiology.

So far we have been considering mainly the large-scale engineering works. What about the workers? During the Proterozoic, a new type of cell evolved, those with nuclei, called the eukaryotes. These are cells that contain structures within them, and other organelles (such as chloroplasts, the green-pigmented bodies that do the work of photosynthesis). Lynn Margulis has taught that these more complex cells are really communities of bacteria that once lived free but now are contained within the outer membrane of one of them. In her book *Early Life* she tells how the presence of oxygen in the Proterozoic set the scene for the appearance of these new and more powerful cells. It was an evolutionary step like that which occurred in the Archean, when the ecosystem of photosynthesizers using carbon dioxide came to equilibrium with the methanogens that returned carbon to the atmosphere as methane and carbon dioxide.

Oxygen opened a giant new niche for organisms that could survive in it and use it. At first these new organisms, who gained energy by combining the organic matter with oxygen, may have existed peacefully with the photosynthesizers, merely eating their debris and dead bodies. But before long there would be consumers, organisms that had learnt to eat fresh food and that grazed the photosynthesizers as they grew. Cells do not have mouths, but they can ingest other cells by enclosing them within a pocket in their membrane, a process called phagocytosis. The pocket becomes part of the cell's interior and dissolves away, leaving the captive entrapped within. Digestion would be the normal fate, but sometimes the roles would be reversed and the ingested organism could be the aggressor. Tubercle and leprosy bacteria do this trick even today, attacking the phagocytes that ingest them instead of succumbing as victims to the phagocytes' powerful digestive system. The results of warfare, however, are rarely genocide; instead war can lead to a peaceful coexistence mutually beneficial to the victim and the aggressor. In this way, the chloroplasts have as ancestors the cyanobacteria of the Archean, and today they power the cell communities of cabbages and redwood trees. Although briefly discussed by nineteenth-century biologists, this powerful association of organelles working within cells in a process of symbiosis, called endosymbiosis, is recognized due to Lynn Margulis more than to anyone else. Endosymbionts enlarged and expanded the possibilities of planetary manipulation by the biota and were a main feature of the history of the Earth during the Proterozoic.

The formation of collectives gives power to the assembly

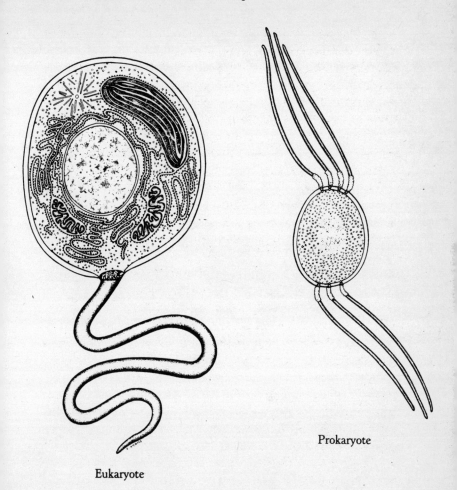

Prokaryote

Eukaryote

5.2 Eukaryotic and prokaryotic cell structures compared. The eukaryotes differ in having membrane-bound organelles which include the nucleus, mitochondria, and chloroplasts. (Drawing by Christie Lyon.)

greater than that possessed by its individual components; but this is never without a price. For the early bacteria (the pro-karyotes), aging was not a problem. They had neither nucleus nor organelles, and they carried their genetic information on a few strands of DNA within the cell membrane (see figure 5.2). Genetic information lost by an individual bacterium during its brief life span was recovered by the exchange of plasmids and other pieces of polymeric software with other organisms. But for the eukaryotic cell with its complex internal organization and organelles (figure 5.2), each carrying a different set of genetic instructions, the loss of some vital item of information by one organelle could mean the death of the cell. The invention of a method and mechanism for the deliberate transfer of information between cells before division greatly reduced the chances of a lethal decision. It was this need that led to the invention of sex. It is much too interesting a story to try to abstract here, and is given in full detail in Lynn Margulis and Dorion Sagan's book *The Origins of Sex*.

An unanswered question about the Proterozoic is, What was the concentration of oxygen? Did it just stay at around 0.1 to 1 percent, or did it rise in concentration to present levels or higher?

Free oxygen comes from two sources: the escape of hydrogen to space and the burial of carbon or sulfur. The sequestering of elemental hydrogen, carbon, or sulfur always leaves free oxygen behind. As we saw in the previous chapter, once oxygen appears in the free state in the air, the escape of hydrogen becomes vanishingly small. This is because only traces of hydrogen or hydrogen-bearing gases like methane can exist in the free state in an oxygen atmosphere. The one exception is water, which cannot be further oxidized, and is confined to the lower atmosphere by the low temperatures at the base of the stratosphere. Quite literally, it is frozen out; and the upper atmosphere contains only a few parts per million of water vapor. The present rate of escape of hydrogen to space is limited by the dryness of the upper air and is only 300,000 tons a year. This is equivalent to just under 3 million tons of water, and would leave behind an excess of 2.5 million tons of oxygen. It sounds a lot, but a loss of water of that rate would have removed less than one percent of the oceans in the age of the Earth.

Once hydrogen loss was reduced to trivial significance, the only way to add more oxygen was to separate carbon and sulfur from combination with oxygen in carbon dioxide and sulfates.

If the separated carbon and sulfur could be buried in the sediments before they had an opportunity to react again with oxygen, a net increment of this gas would be added to the air. This process of separation starts with photosynthesis, which splits carbon dioxide into oxygen, which then enters the air, and into the living and dead parts of the plants and bacteria. Most of this carbonaceous material is recombined with oxygen by consumers, but a little, about 0.1 percent, is buried more or less permanently. Some of the carbon in the sediments is used to reduce sulfate to sulfides. The burial of sulfides also leaves a net increment of oxygen in the air. The carbon and sulfides are buried in the sediments mixed with shales and limestones. The burial can take place in such a way as to form fossil fuels, coal and oil; but these represent only a small proportion of the total carbon and sulfur in the sediments. The burial of all the oxidizable material is like a loan drawn against the oxygen account. So long as it is buried or lost in the Earth's interior, the debt is not presented and free oxygen can remain in circulation in the air.

At present about 100 million tons of carbon are buried each year—equivalent to the release of 266 million tons of free oxygen gas to the air. (This does not mean that oxygen of the atmosphere is increasing, for the increment is all used up by the oxidizable materials released by volcanoes, by weathering, and by processes at the sea floor.) The rate of carbon burial has been constant throughout the history of life on Earth; there is very little difference between the Archean and now. This is curious when you consider that the mass and the activity of the biota may have been less in the Archean. The puzzle can be solved if we remember that because there was only a trace of oxygen present, the proportion of oxic consumers to anaerobes would have been less than now. This means that the methanogens and other organisms of the anoxic sector were digesting nearly all the products of photosynthesis, but buried the same amount of carbon as now. The high rate of photosynthesis today must, in part, be due to the rapid recycling of carbon by the oxygen-breathing consumers. They metabolize 97.5 percent of the products from photosynthesis, leaving only 2.5 percent for the anaerobes. In the Proterozoic, there were consumers present feeding on organic matter and using oxygen to metabolize it. Their activity is likely to have been less than now but greater than in the Archean.

The key point is that oxygen production is determined by the amount of carbon buried, and this in turn depends upon the proportion of the products from photosynthesizers that reach the anoxic sector. Obviously, if the consumers eat all the organic matter there would be none left to be buried and, therefore, no source of oxygen. If we recall that the rate of carbon burial has been more or less constant, then it follows that the input of oxygen from this source has also been constant. In the Archean all this oxygen went to oxidize the reducing substances in, and being added to, the environment, but when free oxygen appeared an increasing proportion of it was used by consumers. The continued existence of the Archean anoxic ecosystems ensured the continuous burial of carbon and a continued input of oxygen to the air. These possibilities are summarized in table 6.1. What, then, determined the oxygen level of the air? Arguing from geophysiology, we can suppose that the inherent toxicity of oxygen is not entirely overcome by the antioxidant systems and by the enzymes of the organisms of the oxic sector; in these circumstances oxygen may set its own limit. Like temperature, it would be an environmental property with a lower and an upper limit for mainstream life. Such properties can be geophysiologically regulated.

Figures 5.3 and 5.4 illustrate how it might have been done in the Proterozoic. Figure 5.3 depicts the effects of oxygen on the growth of the oxic ecosystem and the effect of the size of the oxic ecosystem on oxygen. The solid line is the relationship between a steady level of oxygen and the population of oxygen-using consumers; at low oxygen levels they could not metabolize and at high levels they would be poisoned by oxygen. The dashed line indicates the relationship between the population of the oxic ecosystem and the steady-state level of oxygen; the more photosynthesizers, the more oxygen. The two curves inter-

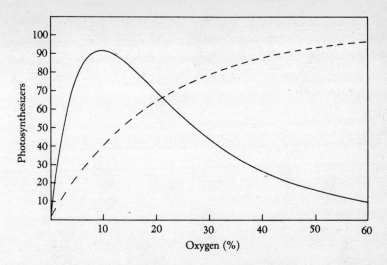

5.3 The effect of oxygen on the growth of organisms (*solid line*) and the effect of the presence of organisms on the abundance of oxygen (*dashed line*). Where the two curves intersect is the level of oxygen at which the system regulates.

sect at an oxygen level that would be kept in homeostasis by the system.

Figure 5.4 illustrates the calculations of a computer model where photosynthesizers, consumers, and anaerobes coexist on a planet before, during, and after the appearance of oxygen. It is assumed that, as on the Earth, the constant burial of carbon and a declining turnover of reducing rocks and gases oxidizes the planet until free oxygen becomes a dominant gas. Thereafter oxygen rises until the geophysiological properties of the system establish a new steady level where the abundance of oxygen is kept constant by a balance between the quantity of carbon buried and the quantity of reducing material exposed. The lower panel illustrates the planetary temperature variations, in comparison with those of a lifeless planet of the same composition. The middle panel shows the variation of the oxygen, carbon dioxide, and methane gases. The upper panel shows the population levels of the different life forms. This model is a linear descendant of the climate models in chapters 2 and 3, where

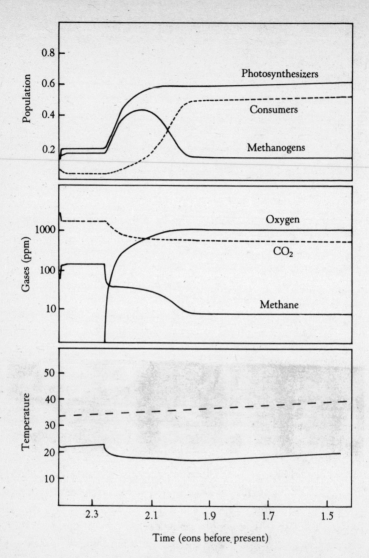

5.4 Model of the transition from the Archean to the Proterozoic. The lower panel shows climate with a lifeless world (*dashed line*) compared with a live world (*solid line*). Note the sudden fall of temperature when oxygen appears. The middle panel shows the abundance of atmospheric gases (carbon dioxide, *dashed line;* oxygen and methane, *solid lines*). The upper panel illustrates the changes in population of the ecosystems as the transition is entered and passed. Note how both photosynthesizers and methanogens increase when oxygen first appears and how methanogens fall back to a steady level when the oxygen-breathing consumers (*dashed line*) become established.

the competitive growth of differently colored daisies is shown to be capable of regulating the temperature of a model planet. It accepts that, in the long term, a constant amount of the carbon produced by photosynthesis is buried and that this source of oxygen is constant. The sink for oxygen would be declining. At the end of the Archean, oxygen rose in abundance. The presence of excess oxygen would have increased the rate of weathering and so increased the supply of nutrients, which in turn would have favored a larger ecosystem. More carbon would have been buried, and the rise in oxygen would have accelerated until toxicity began to set a limit. By this time, the anaerobic sector from which carbon burial takes place would have shrunk to the same size as in the Archean, and oxygen production would again equal oxygen loss by the exposure of oxidizable substances during weathering.

In one sense the oxic ecosystems existed right from the beginning of Gaia; from the moment when the first cyanobacteria converted sunlight into high potential chemical energy and were able to make organic compounds and oxygen from water and carbon dioxide. As the cyanobacteria spread, they would always have occupied a surface position to enjoy and feed on the sunlight. The anoxic systems, whose food was the dead bodies and products of the cyanobacteria, would naturally have existed below the photosynthesizers to take advantage of the conveyance by fallout of the food from above. From the start, there would have been a segregation of these two ecosystems, and a gradient of oxygen concentration declining away from the region of its production.

In the real world, the oxygen cycle cannot be disconnected from the carbon dioxide cycle; as oxygen rises it would be anticipated that carbon dioxide would fall. The carbon dioxide cycle is coupled with the climate, and this in turn affects the growth of both consumers and producers. The environmental feedback from carbon dioxide and climate would further stabilize the system. Once the initial oxygen crisis was over the Proterozoic could have been a comfortable time for Gaia, apart from the persistent annoyance of planetesimals. The natural level of carbon dioxide would have provided a pleasant climate, and no great effort would be needed to regulate it.

A bizarre consequence of the appearance of oxygen was the advent of the world's first nuclear reactors. Nuclear power from its inception has rarely been described publicly except in hyperbole. The impression has been given that to design and construct a nuclear reactor is a feat unique to physical science and engineering creativity. It is chastening to find that, in the Proterozoic, an unassertive community of modest bacteria built a set of nuclear reactors that ran for millions of years.

This extraordinary event occurred 1.8 eons ago at a place now called Oklo in Gabon, Africa, and was discovered quite by accident. At Oklo, there is a mine that supplies uranium mainly for the French nuclear industry. During the 1970s, a shipment of uranium from Oklo was found to be depleted in the fissionable isotope ^{235}U. Natural uranium is always of the same isotopic composition—99.27 percent ^{238}U, 0.72 percent of ^{235}U, and traces of ^{234}U. Only the ^{235}U isotope can take part in the chain reactions necessary for power production or for explosions. Naturally, the fissionable isotope is guarded carefully and its proportion in uranium subjected to thorough and repeated scrutiny. Imagine the shock that must have passed through the French atomic energy agency when it was discovered that the shipment of uranium had a much smaller proportion of ^{235}U than normal. Had some clandestine group in Africa or France found a way to extract the potent fissionable isotope, and were they now storing this for use in terrorist nuclear weapons? Had someone stolen the uranium ore from the mine and substituted spent uranium from a nuclear industry elsewhere? Whatever had happened, a sinister explanation seemed likely. The truth, when it came, was not only a fascinating piece of science but must also have been an immense relief to minds troubled with images of tons of undiluted ^{235}U in the hands of fanatics.

The chemistry of the element uranium is such that it is insoluble in water under oxygen-free conditions, but readily soluble in water in the presence of oxygen. When enough oxygen appeared in the Proterozoic to render the ground water oxidizing, uranium in the rocks began to dissolve and, as the uranyl ion, became one of the many elements present in trace quantities in flowing streams. The strength of the uranium solution would have been at most no more than a few parts per million, and uranium would have been but one of many ions in solution. In the place that is now Oklo such a stream flowed into an algal mat that included microorganisms with a strange capacity to collect and concentrate uranium specifically. They performed their unconscious task so well that eventually enough uranium oxide was deposited in the pure state for a nuclear reaction to start.

When more than a "critical mass" of uranium containing the fissionable isotope is gathered together in one place there is a self-sustaining chain reaction. The fission of uranium atoms sets free neutrons that cause the fission of more uranium atoms and more neutrons and so on. Provided that the number of neutrons produced balances those that escape, or are absorbed by other atoms, the reactor continues. This kind of reactor is not explosive; indeed it is self-regulating. The presence of water, through its ability to slow and reflect neutrons, is an essential feature of the reactor. When the power output increases, water boils away and the nuclear reaction slows down. A nuclear fission reaction is a perverse kind of fire; it burns better when well watered. The Oklo reactors ran gently at the kilowatt-power level for millions of years and used up a fair amount of the natural ^{235}U in doing so.

The presence of the Oklo reactors confirms an oxidizing environment. In the absence of oxygen, uranium is not water soluble. It is just as well that it is not; when life started 3.6 eons back, uranium was much more enriched in the fissile isotope ^{235}U. This isotope decays more rapidly than the common isotope ^{238}U, and at life's beginning the proportion of fissile uranium was not 0.7 percent as now but 33 percent. Uranium so enriched could have been the source of spectacular nuclear fireworks had any bacteria then been unwise enough to concentrate it. This also suggests that the atmosphere was not oxidizing in the early Archean.

Bacteria could not have debated the costs and benefits of nuclear power. The fact that the reactors ran so long and that there was more than one of them suggests that replenishment must have occurred and that the radiation and nuclear waste from the reactor was not a deterrent to that ancient bacterial ecosystem. (The distribution of stable fission products around the reactor site is also valuable evidence to suggest that the problems of nuclear waste disposal now are nowhere near so difficult or dangerous as the feverish pronouncements of the antinuclear movement would suggest.) The Oklo reactors are a splendid example of geophysiological homeostasis. They illustrate how specific minerals can be segregated and concentrated in the pure state—an act of profound negentropy in itself, but also an invaluable subsystem of numerous geophysiological processes. The separation of silica by the diatoms and of calcium carbonate by coccolithophoridons and other living organisms, both in nearly pure form, are such processes and have had a profound effect on the evolution of the Earth.

If some descendant of the alien chemist who visited in the Archean returned in the Proterozoic, it would find an Earth not so different from now. The sky would have been a paler shade of blue with perhaps less cloud cover. On the beach, the sea would be blue-gray, rather than the brown of the Archean. Inland, behind the sand dunes and pebbles, the bacterial mats would lie, enlivened by the origin of certain green and golden yellow algae, protecting the anoxic sector that overlay and kept intact the evaporite beds beneath with their deep layers of lifeless salt, the accumulation of thousands or even millions of years. Out to sea there would again be the rock structures of stromatolite colonies. I wonder if our alien would have observed a remarkable geophysiological property of reefs that has recently been revealed in coral reefs. Satellite photographs showed that the wavelength of ocean waves in the vicinity of the reefs was unusual and unexpected for the surface wind and sea conditions. Subsequent investigation uncovered the remarkable fact that the coral microorganisms secreted a lipid substance that formed a monomolecular layer on the ocean surface and so altered its surface tension as to modify the waves. It is engaging to speculate about the geophysiological development of this remarkable action in microorganisms, and to wonder when it developed and whether it is a mechanism for protecting the reefs from wave damage.

During the Proterozoic, the constant rain of planetesimals continued. As well as numerous smaller ones, there were at least ten that did damage to Gaia comparable in severity to that of a burn affecting 60 percent of the skin area of a human. These events are not in themselves the main interest; we have no detailed information of the dates and consequences of their occurrence. What is interesting is the recovery of the system from these insults. We are tolerably certain that none of them killed Gaia, so that a new Gaia had to be born from the debris. The ability to recover from major perturbations is a test of the health of a geophysiological system; the fact that life persisted and recovered from so many of these catastrophes is more evidence in favor of the existence of a powerful homeostatic system on Earth.

At the beginning of the Proterozoic, the Sun was cooler. The problem facing Gaia was to keep the carbon dioxide greenhouse from collapsing and so causing the Earth to freeze. Without the cooling tendency of life today, the Earth would be uncomfortably hot. It could be said that life is at present keeping the Earth cool by pumping down carbon dioxide. In the middle ages of the Proterozoic, some 1.5 eons ago, the Sun's output was about just right for life and no great effort was needed for thermostasis. The atmospheric carbon dioxide was probably around one percent by volume. It was at a level where physicists and geophysiologists would have no cause for disagreement.

6

Modern Times

I never knew how soothing trees are—many trees and patches of open sunlight, and tree presences—it is almost like having another being.

D. H. LAWRENCE, *Selected Letters*

This chapter is about the period of the Earth's history when living organisms large enough to be seen with unassisted eyes were growing or moving on the land and in the sea. The microorganisms were still there flourishing and still responsible for much of the regulation of the Earth. But the arrival of large, soft-bodied cell communities changed the surface of the Earth and the tempo of life upon it: Plants that could stand erect supported by structures of deep underground roots. Consumers that could travel on the ground and in the air or sea. All these things left fossil remains. Their presence delineates this period called the Phanerozoic, going from the Cambrian some 600 million years ago until the present day. Because we live in it, and because recent historical records are so much more detailed than those of the ancient past, it seems a period well known and familiar. This is an illusion. We know little about the Earth even in our own time. For the Cambrian there are just catalogs of species and rocks. They give some insight into the life of the Earth, but only in the abbreviated way that a telephone book does about the private lives and the economy of a town.

If we take Gaia to be a living organism, the Phanerozoic can be viewed as the most recent stage in her life, and the one she is still in. This may be easier than considering independently the lives of the billions of organisms from which she is made. Getting to know a friend does not usually require a detailed knowledge of her cellular structure. Similarly, geophysiology, concerned with the whole Earth, need not be too confused by the mass of undecomposed detail that lies, like thick layers of fallen leaves, beneath the branches of the tree of science. So let us look at the physiology of Gaia during this period. In an ideal history the description would be of the whole system, but the habit of reduction dies hard. At the present stage of ignorance it is much easier to divide the chapter into parts, each concerned mainly with the regulation of one important chemical element and of the climate.

Geologists see the transition from the Proterozoic to the Phanerozoic as occurring about 570 million years ago. The first organisms that we would recognize as animals with skeletons appeared on Earth somewhat earlier than this. As a geophysiologist I prefer to see this transition as also marked by a change in oxygen abundance, an event not unlike the one that occurred between the Archean and the Proterozoic.

My colleagues have made it very clear to me that what follows about oxygen is speculative and often contrary to conventional wisdom. I have included it in spite of their protest because it illustrates a view of the evolution of free oxygen in the light of Gaia theory. Whether it is right or wrong seems to me less important than its value in stimulating new experiments and measurement.

So let us consider oxygen. This gas comes from the use of sunlight by the green chloroplasts within cells to convert carbon dioxide and water into free oxygen and the biochemicals from which they are made. Most of the oxygen is used up again by the consumers who eat the plants and algae, oxidize the food, and return carbon dioxide to the air and the sea. From the beginning the producers, the photosynthesizers, have had a love-hate relationship with the consumers. Producers do not care to be eaten, but the presence of the consumers is essential for their health and that of the larger organism they constitute. When plants and animals appeared, the fine details of this constructive aggression became visible. The plants were seen to possess poisons, spines, and stings; and the animals and microorganisms were obliged to develop new techniques for grazing. A balance is always struck because, without the consumers, the survival of the plants and algae would be threatened. There is only a few years' supply of carbon dioxide in the air. The removal of consumers from the scene would be disastrous for plants, and within a short time span. Not only would there be too little carbon dioxide for photosynthesis, but there would be major climate changes as the gases of the atmosphere and the albedo of the Earth responded to the demise of the plants. Not least, the intricate recycling of nutrients and gardening of the soil would cease. On a human time scale the coexistence of consumers and producers could be compared with the long peace that has reigned between the hostile yet mutually dependent superpowers.

Oxygen is also used up in its reaction with, for example, the sulfur gases emitted by volcanoes, or the reducing chemicals in the igneous rocks that solidify from the magma emerging from below the sea floor. Oxygen is kept at a constant level by the burial of a small proportion of the photosynthetic carbon, about 0.1 percent, just enough to equal the losses. We know that the level of oxygen must have changed at the end of the Proterozoic, because of the new forms of life that appeared.

When the organisms were mostly living in water, or as colonies of algal mats on the surface of the land, the upper limit of oxygen would have been set by its toxicity. For such ecosystems, fires are less a problem than they are to standing vegetation. They could have tolerated an atmosphere containing as much as 40 percent oxygen, provided that the extra atmospheric pres-

sure did not so exacerbate the gaseous greenhouse as to lead
to an intolerably hot climate.

However, the free-swimming eukaryotes that appeared in the
early Proterozoic would not have required much oxygen since
the gas could diffuse easily across the small distance between
the walls of their microscopic cells; as little as 0.1 percent in
the atmosphere may have been sufficient. The larger organisms
that appeared in the Phanerozoic, such as the dinosaurs, which
were composed of massive volumes of cells in juxtaposition,
could have existed only in a richer oxygen environment. This
is especially true where there was a need for a greater power
output during swimming. Even today, with oxygen at 21 percent,
our muscles cannot be supplied with sufficient oxygen at maximum power output; a backup temporary power supply, called
glycolysis, operates when we run as fast as we can. Peter Ho-
chachka, in an unusual book called *Living Without Oxygen,*
describes the intricate mechanisms by which large animals cope
with the problem of power production in a world which, for
them, can be limited in its oxygen supply. An example of this
size effect is illustrated by the poison carbon monoxide. For
animals as large as ourselves, carbon monoxide is inescapably
deadly. It kills by preventing the red blood cells from conveying
oxygen to our tissues. A smaller animal, the mouse, can survive
the complete saturation of its blood with carbon monoxide. It
survives the poison because enough oxygen can diffuse to its
tissues from the skin and from the surface of the lungs.

There has to be an upper limit of oxygen concentration at
which these large animals can live because of the toxic effects
of this gas. We are so accustomed to think of oxygen as life-
saving and essential that we ignore its potent toxicity. Oxidative
metabolism, the extraction of energy from food through its reaction with oxygen, is inevitably accompanied by the escape of
highly poisonous intermediates within the cell. A substance
like the hydroxyl radical is such a powerful oxidant that were
it present as a gas at the same concentration as oxygen, almost
anything flammable would instantly burst into flame. It reacts

with methane at room temperature, whereas free oxygen does not until nearly 600°C. Other undesirable products from oxygen are hydrogen peroxide, the superoxide ion, and oxygen atoms. Living cells have developed mechanisms to detoxify all those products: Enzymes, such as catalase, that decompose hydrogen peroxide to oxygen and water, and the superoxide dismutase, which converts the malign superoxide ion to harmless products. Antioxidants, such as tocopherol, that mop up hydroxyl radicals. We and other animals alive today, from the largest to the smallest, owe our life spans to this system of chemical protection developed by our distant bacterial ancestors. If there is no great excess of oxygen, its toxicity can be contained.

Why did the level of oxygen rise? At the end of the Archean, the supply of reductants—sulfides and ferrous iron—of the early Earth became insufficient to match the flux of oxygen coming from the burial of carbon, and the oxygen increased. It reached a low steady state in the early Proterozoic, much less than in the present atmosphere, representing a balance between the needs of early consumers and the toxicity of oxygen to the early photosynthesizers. There is no similar clear-cut event in the Proterozoic corresponding to the appearance of oxygen at the end of the Archean (see table 6.1). We do not know why the level of oxygen began to rise again, although Robert Garrels proposes that it was associated with the development of bacteria that reduce sulfates. This would have led to the burial of more of the products from the photosynthesizers, as sulfur or sulfides,

Table 6.1 OXYGEN SOURCES AND SINKS

| | | | Sinks | |
PERIOD	ABUNDANCE	SOURCES	CONSUMERS	ROCKS
Archean	10^{-7} to 10^{-5}	10	1.0	9.0
Proterozoic	0.01 to 0.1	30	29.8	0.2
Phanerozoic	0.21	100	99.9	0.1

NOTE: The abundances of oxygen are expressed as "mixing ratios," that is, proportions of the total atmosphere. The sources and sinks are the quantities of oxygen, Giga-tons per year, flowing to and from the atmosphere. The present photosynthetic flux is about 100 in these units.

so leaving behind an excess of oxygen in the air. However it happened, the reactions of this free oxygen with other elements such as carbon and sulfur would release acids into the air, and these would increase the weathering of crustal rocks so that more nutrients were released, leading to a greater abundance of living organisms. The positive feedback on the growth of oxygen would continue until the disadvantages of its presence overcame the benefits. Rather like the growth of car population in some cities, it continues until movement is choked by its presence.

At some time in this period organisms began synthesizing, on a large scale, the precursors of those enigmatic substances, lignins and humic acids. It may have been the result of the invention of some new antioxidants. The precursors of lignins are phenols, well known to react vigorously with hydroxyl radicals. A typical member of this class of acid substances is coniferyl alcohol: when it reacts with hydroxyl it produces lignin, a carbon-containing polymer that has great chemical stability and a resistance to biodegradation. Because of these properties lignin would, if made in quantity, increase the rate of carbon burial, and thus the rate of oxygen production. In a geophysiological fashion, lignin has turned out to be a structural material as important for land plants as the bioceramics of bone and shells are for animals. Just as calcite deposition in cells may have originally been part of a device to lower the concentration of the toxic calcium in the cell fluids, so lignin production may have initially come from a method of detoxifying oxygen. Both of these materials enabled the construction of vast cell communities of a new kind. At first in the oceans, but now in the living organisms we recognize as plants and animals.

The model of the evolution of oxygen and carbon dioxide regulation, illustrated in figure 5.4, can be extended to the present day. But it is unable, as it stands, to account for the precise regulation of oxygen observed for the past several hundred million years. Oxygen has been constant at 21 percent by volume in the Phanerozoic. The evidence of this constant high concentration is the presence in the sediments of layers containing charcoal.

These can be found as far back as 200 million years. The presence of charcoal implies fires, probably forest fires. This sets sharp limits on the atmospheric oxygen abundance. My colleague, Andrew Watson, showed that fires cannot be started, even in dry twigs, when oxygen is below 15 percent; above 25 percent oxygen, fires are so fierce that even the damp wood of a tropical rain forest would burn in an awesome conflagration. Below 15 percent there could be no charcoal; above 25 percent no forests. Oxygen is 21 percent, close to the mean between these limits.

It might be that fires themselves are the regulator of oxygen. There is no shortage of lightning strikes for their ignition. If fires are the regulator it cannot be a simple relationship. Oxygen in the air comes from the burial of carbon. Consumers are efficient, and only about 2 percent of carbon photosynthesized reaches the sediments, where a further 95 percent is returned to the oxidized environment as methane. So only one part in a thousand of the carbon fixed by the plants is buried deep. Combustion, on the other hand, is inefficient. As any charcoal maker will tell you, up to 70 percent of the carbon of wood can remain from a controlled combustion. Fires, therefore, would lead to the burial of much more carbon, because charcoal is entirely resistant to biological degradation. Paradoxically, then, fires lead to more oxygen in the long run. If this grim scenario is followed to a conclusion there would at first be a positive feedback on oxygen, but soon the forests would be so devastated that carbon production would fall to the point where oxygen was near or below its present level. The cycle would then repeat. It is true that the layers of charcoal present in the sediments suggest recurrent fires, but the proportion of buried carbon existing as charcoal is much too small to account for such a cycle.

A more subtle regulation involving fire would come from the use of fires by certain species of tree as a weapon to sustain its possession of territory. The conifers and eucalyptus trees have both independently evolved to produce on the forest floor a highly flammable detritus: piles of kindling rich with resin and terpenes that ignite and burn fiercely at a lightning stroke. This contrived form of fire does not damage the tall trees themselves, but is death to competing species such as oaks. Further-more, these fires leave little charcoal; combustion is nearly com-plete. So developed is the fire ecology of forests that some conifer species require the heat of fire to release their seeds from the seed capsules. The regulation of oxygen so precisely at the con-venient level of 21 percent does suggest that the large plants, flammable and nonflammable, who are the victims and beneficia-ries both, play a key part. I can't help wondering if those flamma-ble trees that use fire ecology also carry less lignin than other vegetation. If so, they would be a lesser source of buried carbon and so serve to regulate oxygen at a level where fires did take place but not so fiercely as to do more harm than good.

The separate discussion of oxygen is justified by its historical significance; it is almost as if oxygen were the conductor who led the players in their evolutionary orchestra. But we need remember that in Gaia the evolution of the organisms and their environment constitute a single and inseparable process. In addi-tion the cycles of all the elements that make up Gaia are closely coupled among themselves, as well as with the species of the organisms. Attempts to describe the role of each of these parts of the system separately are crippling to insight but made neces-sary by the unavoidable use of the linear form of written expres-sion. With this thought in mind, and remembering that the geophysiology of oxygen and carbon cannot be separated, let us now look at carbon dioxide.

In modern times, carbon dioxide is a mere trace gas in the atmosphere compared with its dominance on the other terrestrial planets or with the abundant gases of Earth, oxygen and nitrogen. Carbon dioxide is at a bare 340 parts per million by volume now. The early Earth when life began is likely to have had 1,000 times as much carbon dioxide. Venus now has 300,000 times as much; and even Mars, with much of its carbon dioxide frozen in the surface, has 20 times as much. James Walker and his colleagues tried to explain the low carbon dioxide of the Earth by a simple geochemical argument. Their model was based on the facts that the only source of the gas is volcanic emission and the only sink its reaction with calcium silicate rock. In their world, life played no part in the regulation of carbon dioxide. As the Sun warmed, two processes took place. The first was an increase in the rate of evaporation of water from the sea and, hence, rainfall; the second, an increase in the rate of the reaction of carbon dioxide with the rocks. Together, these processes would increase the rate of weathering of the rocks and so decrease the carbon dioxide. The net effect would be a negative feedback on the temperature rise as the solar output increased. Unfortunately, this imaginative and plausible model could not explain the facts. The carbon dioxide it predicted for the present was about 100 times more than it is observed to be.

James Walker's model can be brought to life by including within it living organisms. If the soil of a well-vegetated region almost anywhere on Earth is examined, the carbon dioxide content is between 10 and 40 times higher than the atmosphere. What is happening is that living organisms act like a giant pump. They continuously remove carbon dioxide from the air and conduct it deep into the soil where it can react with the rock particles and be removed. Consider a tree. In its lifetime it deposits tons of carbon gathered from the air into its roots, some carbon dioxide escapes by root respiration during its lifetime, and when the tree dies the carbon of the roots is oxidized by consumers, releasing carbon dioxide deep in the soil. In one way or another living organisms on the land are engaged in the business of pumping carbon dioxide from the air into the ground. There it comes into contact with, and reacts with, the calcium silicate of the rocks to form calcium carbonate and silicic acid. These move with the ground water until it enters the streams and rivers, on their way to the sea. In the sea, the marine organisms continue the burial process by sequestering silicic acid and calcium bicarbonate to form their shells. In the continuous rain of microscopic sea shells, the products of rock weathering—sedimented limestone and silica—are buried on the sea floor and eventually subducted by the movements of plate tectonics. Were life not present, the carbon dioxide from the atmosphere would have to reach the calcium silicate of the rocks by slow inorganic processes like diffusion. To sustain the same soil carbon dioxide as now the atmospheric concentration would have to be even higher, perhaps as much as 3 percent. This is why the Walker model will not work.

Considered in this way, we have an explanation for the low carbon dioxide of today's Earth. This great geophysiological mechanism has served since life began as one part of climate regulation. But as the Sun grows hotter, it can have little chance of continuing to keep our planet cool. There is an inverse relationship between the abundance of carbon dioxide and the abundance of vegetation. Assuming that the health of Gaia is measured by the abundance of life, then periods of health will be at times of low carbon dioxide. During the normal healthy state of Gaia, with the comfortable coolness of a glaciation, carbon dioxide is a bare 180 parts per million by volume—uncomfortably close to the lower limit for the growth of plants. Not surprising is the emergence in the Miocene, some 10 million years ago, of a new type of green plant able to grow at lower carbon dioxide concentrations. These plants have a different biochemistry and are called C4 plants to distinguish them from the mainstream C3 plants. The names C3 and C4 come from a difference in the metabolism of carbon compounds in these two types of plant: the C4 plants are able to photosynthesize at much lower carbon dioxide levels than the older C3 plants. The new C4 plants include some, but not all grasses, whereas trees and broadleaved plants generally use the C3 cycle. Eventually, and probably suddenly, these new plants will take over and run an even lower carbon dioxide atmosphere to compensate for the increasing solar heat. But it will serve only temporarily, because in as short a time as 100 million years, assuming nothing else had changed, the Sun will have warmed up enough to require a zero carbon dioxide atmosphere to keep the present temperature. As we shall shortly see, there are other cooling mechanisms that could come into play. Also a different ecosystem could evolve that was comfortable with a global mean temperature even as high as 40°C. The carbon dioxide crisis is serious but not necessarily life-threatening to Gaia.

If I am right that the glacial cool is the preferred state of Gaia, then the interglacials like the present one represent some temporary failure of regulation, a fevered state of the planet for the present ecosystem. How do they come about?

Active systems of regulation or control are well known to exhibit instability when close to the limit of their operating range. This can be clearly seen in the Daisyworld model in figure 3.6 where, as the star warming the imaginary planet grows hotter, the effects of a cyclical plague affecting the plants appear in an amplified form as cyclical fluctuations of temperature until the system fails from overheating. We do not yet know the cause of the glaciations, but we do know that they are a periodic phenomenon, synchronized with small variations in the amount of solar radiation reaching the Earth and with long-term variations in the Earth's inclination and orbit. This astrophysical link between glaciation and the Earth's orbit and inclination was proposed by a Yugoslavian, Milutin Milankovich. The magnitude of the change in warmth received from the Sun is not in itself enough to account for the range of temperature between the glacials and interglacials, but it could be the trigger synchronizing the change from one state to another. According to a Japanese physicist, Shigeru Moriyama, the mathematical analysis of the periodicity of the Earth's mean temperature during the past million years is more consistent with an internal oscillation, triggered externally, than with an oscillation that was free running, or simply a response to the changes in radiant energy received from the Sun.

Geophysiology suggests that, to regulate the climate in face of increasing heat from the Sun, glacials are the normal state and the interglacials, like now, are the pathological one. Thinking this way, the low carbon dioxide during the glacials can be explained by the presence of a larger or more efficient biota. There must have been *more* living organisms on Earth; how else could the carbon dioxide have been so low? If more organisms were doing the pumping, where were they? At first thought it might seem that the ice sheets would leave less room for life as it covered much of what is now, or was before humans, forested land. However, as water was used to form the land-based glaciers, the level of the sea could have fallen by some 100 meters, exposing vast areas of rich and fertile soil on the continental shelves. A glance at a map of the continental shelves reveals that much of the new land would have been in the humid tropics, such as in present-day Southeast Asia. It could have covered an area comparable with that of Africa now, and could have supported tropical forests.

Such a world is inherently unstable. If a warming trend, as by the Milankovich effect, led to a decrease of land area, then increased carbon dioxide together with the geophysical feedback of a diminution in the area of reflective ice and snow cover would lead to a runaway rise of both temperature and carbon dioxide. The system would also be unstable in a biological sense. Close to the lower limit of carbon dioxide for photosynthesis there would have been intense selection pressure for plants to emerge that could live at even lower carbon dioxide. There are other critical events that could precipitate a rise of carbon dioxide and temperature. One that comes to mind is some effect connected with the increase of salt in the oceans as water froze to form ice. Acid rain from the sulfur emitted by the marine algae as a result of excess salinity (or a failure of the supply of sulfur volatiles from marine biota, which could lead to a decline of land plants by depriving them of an essential element) could be another. A decrease of cloud cover and planetary albedo is yet another. The cycles of the ice ages are known. Figure 6.1 illustrates the time history of temperature during the past million years.

We also need to take into account regional processes that may oppose the general tendency for cooling. In the northern temperate regions the great conifer forests are dark in color and easily shake off or shed the white snow that falls on them in winter. The length of the winter season must be considerably reduced by their presence. The late winter sunshine at continental latitudes greater than 50° is not powerful enough to melt fresh snow; the whiteness of it reflects the radiant energy skywards. Dark pine trees, though, absorb the sunlight and warm

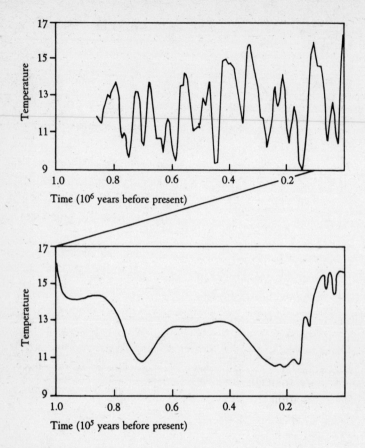

6.1 Temperature history of the recent series of glaciations. (After S. W. Matthews.)

not just the trees themselves but the region. Once the snow has melted then even the bare ground can absorb sunlight, making it warm enough for seeds to germinate and letting the spring commence.

The circularity of explanations of physiological control systems makes it difficult to choose a point of entry. Which came first, the low carbon dioxide and dense cloud cover, or the low temperature? This question, like that about the priority of chickens and eggs, could be pointless. Let us look instead at a recent evolutionary development, the emergence of the C4 plants that are able to grow at lower concentrations of carbon dioxide than the older C3 plants. These C4 plants could be both the result of the glaciations and an encouragement for further glacial periods. Now there is ample carbon dioxide for all plants, so there is not much competition between C3 and C4 plants for habitats, except through the agency of humans who, in agriculture, remove the older C3 plants and replace them by wheat, rice, bamboo, sugar cane, and so on, many of which are C4 plants. During a glaciation, when the carbon dioxide is near the lower limit tolerable for C3 plants, the advantages of the C4 metabolism begins to tip the balance in their favor.

The human propensity to interfere was the plot of a doom scenario in my first Gaia book. The central character was an earnest, well-meaning agricultural biologist, Dr. Intensli Eeger. He succeeded, where all other hazards had failed, in eliminating all life by his meddling. He developed, using genetic engineering, a combined nitrogen-phosphorus fixing microorganism. It was intended to improve the yield of rice grown in the humid tropics so that the hunger of the Third World would at last be overcome. Unfortunately, his organism found a free-living unicellular alga much more to its liking than rice plants. So successful was this combination that it conquered the world. It was a Pyrrhic victory, because the bicultural world of the algal-bacterial combination could not, on its own, maintain planetary homeostasis.

I have had a certain guilt about ascribing, even to a fictional character, so awful a punishment for meddling, and it seems only fair to give him a second chance. This time he uses his impressive skill to develop a new form of tree starting with wild oats, one that would operate on the C4 cycle and grow vigorously in the humid tropics. It would have a rich sap, a delicious fruit full of vitamins and nutrients, and an ability to grow well in arid areas. Its plantations could reverse the spread of desert.

The replacement of much of the humid tropical forests with *Avena eegeriansis* at first gave the impression that the bad days of environmental degradation were over. Lush plantations were sprouting everywhere, greening the Sahel and bringing back rain to regions that had been desert for thousands of years. Under the shade of the new trees, the complex tropical ecosystems began to return. Soon it was noticed that the carbon dioxide greenhouse problem was abating; the lush growth of the trees had so increased the rate of carbon dioxide uptake by the soil that the sink was now larger than the source. Some scientists, though, were commenting that cloud cover, and, hence, albedo had increased. There was a fierce scientific debate. In line with current thinking, and encouraged by the generous supply of research funds, theorists blamed the increased cloudiness on the activities of the chemical and nuclear industries.

Soon the winter snow was lingering in Moscow, Boston, Chicago, Bonn, and Beijing until May; further north it was snowbound year round. Nuclear power stations and the chlorofluorocarbon industry were closed down. But, faster than the great urban populations of the Northern Hemisphere could grasp, the world would be deep into the next and greatest glaciation. Gaia would breathe free again, cool and comfortable at a total atmospheric carbon dioxide of 100 parts per million. It would not be long, in Gaia's terms, before the oceans receded from the vast areas of continental shelf. Australia and Papua New Guinea would once again be joined by land covered with an ever-extending forest. The lands and cities of the superpowers of yesterday would nearly all be buried under the glaciers. C4 plants would have taken over, with the help of humankind, and liberated Gaia for the start of another long period of homeostasis—an ice age to last for millions of years, not just hundreds of thousands.

This is an unlikely story, but it does serve to illustrate the way that a punctuation can happen as a result of a change in dominant species. We might be the highest form of animal life, but without doubt trees are the highest form of plant life. A fully developed C4 tree might be formidable competition for the forest trees we now know. Dr. Eeger would have redeemed himself and led humans back into a seemly existence within Gaia.

In living organisms, the element sulfur is widely used in structures and functions. So next I would like to explain how information gathered in the past decade has enlarged our understanding of the physiological role of sulfur in Gaia.

In the summer of 1971 I attended a Gordon Conference held in New Hampton School in the town of the same name in New Hampshire. The title of the conference was "Environmental Science: Air," and the chairman James Lodge, an atmospheric chemist and a friend. It is no small tribute to his powers of organization that this conference could be said to have marked the start of a deep new interest in the atmosphere that has continued to this day.

It was there that I first presented experimental measurements of the halocarbons and sulfur gases in the air. I also learnt that the conventional wisdom about the natural cycle of sulfur was that it required large quantities of hydrogen sulfide to be emitted from the oceans to make up for the losses of sulfur, as the sulfate ion, in the run-off of rivers. Without some return of sulfur, the land organisms would soon have been starved of this essential element. I knew from Professor Frederick Challenger's work at the University of Leeds in the 1950s that many marine organisms emitted sulfur as the gaseous compound, dimethyl sulfide. I also knew, as a one-time chemist, that hydrogen sulfide was rapidly oxidized in water containing dissolved oxygen, and that it stunk. It seemed to me that, on both these grounds, it could not be the major carrier of sulfur from the ocean to the land. On the other hand, that elusive smell of the sea is much like that of dilute dimethyl sulfide. Indeed, once you have smelt this gas, pleasant when diluted, it is recognizable ever after as a significant component of the aroma of fresh fish straight from the sea. It is not part of the smell of fresh fresh-water fish.

When I returned home to England I thought that it might be a good idea to go by ship from the Northern Hemisphere to the Southern Hemisphere, measuring the sulfur-carrying gases in the air and the sea to try to find out if dimethyl sulfide were indeed the carrier of sulfur in the natural world. I also wanted to take the opportunity to measure the halocarbon gases, such as are used in aerosol sprays, in the hope that these effec- tively "labeled" the air and would allow us to observe its move- ment over the oceans. This was to be the last occasion that I applied for research funds through the regular system of writing a proposal and submitting it to a funding agency. What I sought was a small grant, no more than a few hundred pounds, to make some apparatus and take it by ship from the Northern Hemisphere to the Southern, measuring the gases each day the ship sailed. I should have known better. Both proposals were rejected. To the peer reviewers it was pointless to look for dimethyl sulfide, since it was known that the missing sulfur flux was conveyed by hydrogen sulfide. The second proposal, to look for halocarbons, was rejected as frivolous because it was "obvious" that no apparatus existed sensitive enough to measure the few parts per trillion of chlorofluorocarbons I was proposing to seek.

I was lucky in being independent. All that I needed for ap- proval to make the voyage was the agreement of my wife Helen, whose housekeeping funds would be somewhat diminished by the cost of the research. She did not share the opinions of my "peers." I made a simple gas chromatograph (shown in figure 6.2) whose total cost could not have been more than a few tens of pounds. Some kindly civil servants of the Natural Environ- ment Research Council, who also disagreed with their panel of academic advisers, provided my travel and subsistence ex- penses from a discretionary fund. I traveled on a research ship, the RV *Shackleton,* on its journey from Wales to Antarctica and back. I returned from Montevideo after three weeks on the ship, sadly all the time I could afford; but a fellow voyaging scientist, Roger Wade, kindly continued the measurements when the ship was in Antarctica. My colleague, Robert Maggs, flew out to Montevideo in the spring of 1972 to complete the run home across the equator to Britain. The measurements made on this voyage were reported in three small papers in *Nature.* The first reported the halocarbon measurements, which showed that the chlorofluorocarbons were persistent and long-lived in

6.2 Homemade apparatus used to measure gases, in the sea and air, aboard the RV *Shackleton* on its voyage from Britain to Antarctica and back in 1971 to 1972.

the Earth's atmosphere, and that two other halocarbon gases, carbon tetrachloride and methyl iodide, were to be found wherever the ship sailed. These findings led to among other things, the "ozone war" and to the disbursement of an ocean of research funds, recommended by the same committees that had rejected the first applications. Speculations about the threat to "the Earth's fragile shield," the ozone layer, were more plausible than the idea of a voyage of discovery stimulated by no more than the curiosity of an individual scientist.

The second and third papers on the sulfur gases reported the ubiquitous presence of dimethyl sulfide and carbon disulfide in the oceans. These findings were, apart from the pioneering calculations of the fluxes by Peter Liss of the University of

East Anglia, largely ignored—until M. O. Andreae showed by his careful and extensive measurements of the oceanic sulfur gases, in the early 1980s, that the output of dimethyl sulfide from the oceans was indeed sufficient to justify its role as the major carrier of the element sulfur from the sea to the land.

Dimethyl sulfide would not have been sought as a candidate chemical transporter had it not been for the stimulus of Gaia theory that required the presence of geophysiological mechanisms for such transfers. But what on earth, you may ask, could be the mechanism? Why should marine algae out in the open oceans care a fig for the health and well-being of trees, giraffes, and humans on the land surfaces? How could such an amazing altruism evolve through natural selection?

The answer is not yet known in detail, but we have a glimpse of how it might have evolved from the properties of a strange compound called dimethylsulfonio propionate. This substance is what organic chemists call a betaine, after the discovery long ago of a similar compound, trimethylammonio acetate or betaine, first isolated from beets. The importance of betaines for the health of marine organisms living in a salty environment was discovered by A. Vairavamurthy and his colleagues. Betaines are electrically neutral salts. They carry a positive charge, associated with the sulfur or nitrogen, and a negative charge, associated with the propionic acid ion, on the same molecule. In an ordinary salt, such as sodium chloride, solution in water separates the charges, which become independent free-floating ions. As we saw in the preceding chapter, marine life lives near the limit of tolerable salt concentration. Salt concentrations above 0.8 molar for sodium chloride are toxic, but this does not apply to betaines. The internal neutralization of their ionic charges renders them nontoxic as salts, and they act in a cell like sugars, glycerol, and the other neutral solutes. Cells that are able to substitute a large proportion of betaine for salt are at an advantage.

I wonder if some time, long ago, marine algae were left by the ebbing tide on some ancient beach. The sunlight would soon dry them. As water evaporated from their cells, the internal salt concentration would rise above the lethal limit, and they would die. In the way of evolution, those algae that had present in their cells neutral solutes like the betaines would be less damaged by desiccation and would tend to leave more progeny. In time, the synthesis of betaines would be common among marine algae. Sulfur is plentiful in the sea, whereas nitrogen is often scarce. On the land the reverse is true. This may be why dimethylsulfonio propionate was the chosen betaine rather than the nitrogen betaine of beets and other land plants. (Incidentally beets, also, are able to deal with high concentrations of salt.) This may not be the whole explanation of the presence of dimethylsulfonio propionate as a prominent algal betaine, but there is no doubt that algae that contain it are the source of dimethyl sulfide. When the algae die or are eaten, the sulfur betaine decomposes easily to yield the acrylic acid ion and dimethyl sulfide. Algae that were prone to being left high and dry on the beach would, therefore, have evolved this sulfur gas, and onshore breezes would have carried it inland where atmospheric reactions would slowly decompose it and deposit sulfur as sulfate and methanesulfonate on the ground. Sulfur is scarce on the land and this new source could have enhanced the growth of land plants. The increased growth would increase rock weathering and so increase the flow of nutrients to the ocean. It is not difficult to explain the mutual extension of the land-based ecosystems from the supply of sulfur and of the sea-based ecosystems from the increased flux of nutrients. By this, or some similar series of small steps, the intricate geophysiological regulation systems evolve. They do so without foresight or planning, and without breaking the rules of Darwinian natural selection.

Before leaving the beach, so to speak, I have wondered also about the widespread production of methyl iodide by marine plants. Unlike the innocuous dimethyl sulfide, this compound is toxic. It is a mutagen and a carcinogen. The first stimulus for its production may have been as an antibiotic to help the algae to compete, or to discourage predators. The release of methyl iodide to the air from the sea is an essential mechanism for the maintenance of a continuous supply of iodine, an element that is vital for land organisms. It might be worth investigating the possibility that a specific betaine, methyliodonio propionate, exists in large algae such as the brown seaweed, *Laminaria*, which are a strong source of methyl iodide. If it does, then it suggests a common link with the sulfur betaine story.

But there is more to the sulfur and iodine cycles than just the recycling of nutritious elements. The Alaskan geophysicist, Glen Shaw, had a stimulating idea for an efficient geophysiological climate control system. Knowing that a small (in Earthly terms) quantity of sulfur in the stratosphere could profoundly affect the climate, he proposed that the emission of sulfur gases by marine organisms was the most efficient method of climate control. There is a fair body of evidence to suggest that major volcanic eruptions are followed by a global fall in mean surface temperatures. The volcanic gases injected into the stratosphere by the eruption include sulfur dioxide and hydrogen sulfide. (The volcanic cloud also contains an aerosol of solid material, but this soon settles downward.) The sulfur gases remaining in the stratosphere oxidize and, with the water vapors present there, form submicroscopic droplets of sulfuric acid. Because they are so small, they settle only slowly and may persist for several years. These droplets form a white haze in the stratosphere that returns to space the sunlight that might otherwise warm the Earth. Between eruptions, there remains a background of sulfuric acid droplets that are continuously formed from the oxidation of sulfur gases from living organisms. The most important of these are carbonyl sulfide and carbon disulfide. They are minor emissions compared with that of dimethyl sulfide, but in the lower atmosphere they are only slowly oxidized (carbonyl sulfide is especially slow), and persist long enough to enter the stratosphere and be oxidized there. Glen Shaw's proposal was that global overheating could be offset by marine life increasing its output of carbonyl sulfide and carbon disulfide, leading to a thickening of the haze of sulfuric acid droplets in the stratosphere and so to a cooling of the Earth. This may indeed be one of several available geophysiological mechanisms for climate regulation. But it set my colleagues thinking of what might be a much more potent use of sulfur gases for the same end.

During extensive investigations of the world oceans, M. O. Andreae has shown that marine organisms emit vast quantities of dimethyl sulfide. These emissions are particularly marked over the "desert" areas of the open oceans far away from the continental shelves. This finding led the meteorologists Robert Charlson and Stephen Warren to propose that the rapid oxidation of dimethyl sulfide in the air over the ocean could be the source of the nuclei needed for the condensation of water vapors to form clouds. Small droplets of sulfuric acid are ideal for this purpose, and over the open oceans there is no other significant source of condensation nuclei from which to form clouds. The aerosol of sea salt which might be thought to nucleate cloud droplets is much less efficient than are the microdroplets of the sulfur acids. The oceans cover about two-thirds of the Earth's surface and their color is a dark blue. Anything that affected the cloud cover over the oceans could powerfully affect the climate of the Earth. In a joint paper, the four of us have reported calculations to estimate the effect that the present natural emissions of dimethyl sulfide could have; these suggest that it is comparable in magnitude with that of the carbon dioxide greenhouse, but in opposition to it.

We have shown the possibility of a powerful link between the growth of algae on the ocean surface and the climate. As a geophysiologist I would further ask: Could these processes serve as a significant part of a responsive climate regulation system? And if so, how did this system evolve? We may also need to take into account the iodine cycle, because the oxidation of dimethyl sulfide in the marine atmosphere is catalyzed by iodine compounds. The production of methyl iodide by the algae may also be a part of this system of climate control.

The sites we are proposing for cloud regulation by sulfur emission are the open desert areas of the tropical oceans, about 40 percent of the surface area of the Earth. These regions are low in productivity compared with the continental shelves and inshore waters. They are bare of life, like the great land deserts that span the 30° latitudes north and south of the equator. On land, it is a lack of water that makes the desert; on the oceans it is a lack of nutrients, particularly nitrogen. What are these ocean deserts like? Their waters are clear and blue and, like land deserts, they are by no means devoid of life. One of these deserts is the Sargasso Sea. I recall reading when I was a boy an adventure story about the perils faced on a sailing ship trapped in the dense entangling weed of the Sargasso Sea. When I passed right through that region in 1973 aboard the German research ship *Meteor,* I was amazed at the difference between the reality and my recollections of the story. There was floating weed, but no more than well-dispersed thin strands of bladder wrack—the ocean equivalent of sagebrush in an arid desert, and no more impediment to the motion of the ship than would be the sagebrush to walking across the desert floor.

The algae at the surface of these ocean deserts do not produce the precursors of cloud condensation nuclei for our benefit nor as a part of some grand design to keep the planet cool. The process must have its origins in the local environmental effects of algal biochemistry. I have discussed the possibility that the production of the sulfur betaine, dimethylsulfonio propionate, may have been a cellular response to salt stress. Although it may have been discovered by marine algae drying out on the shore, successful inventions tend to spread. The concentration of salt in the sea is always uncomfortably high for living organisms. For the unicellular or small-floating organisms, unable to regulate their internal salt by osmotic pressure, synthesizing betaines may have been the cheapest way, in terms of energy, of achieving a low-salt interior. Again dimethylsulfonio propionate would have been the natural choice, because sulfur (in the convenient form of the sulfate ion) is abundant, whereas nitrogen is not. The dimethylsulfonio propionate persists in the cells of the algae during their lifetime, but when they die or are eaten it disperses in the ocean, where it slowly decomposes to yield dimethyl sulfide. Both of these compounds are consumed by other organisms, but there is a steady flux of dimethyl sulfide to the air. In the air, the gas is rapidly oxidized by the ubiquitous hydroxyl radicals until nearly all is converted to sulfuric and methanesulfonic acids. The vapors of these acids are carried aloft by the motions of the air until they reach the heights supersaturated with water vapor, where they act as cloud droplet nuclei.

The escape of dimethyl sulfide to the air can bring to the algae inadvertent benefits. The extra cloud cover from the presence of sulfuric acid nuclei changes the local weather. Timothy Jickells of the University of East Anglia has drawn my attention to the fact that clouds over the ocean increase wind velocity, and stir the surface waters, mixing in the nutrient-rich layers beneath the depleted photosynthesizing zone. This is an effective reward for the production of cloud condensation nuclei, and has just been confirmed by the work of the meterologist, John Woods. I doubt if the fresh water of the rain assists much with the salt-stress problem of the algae, but it is no disadvantage. In some regions of the sea, the air carries an aerosol of dust particles blown from the continents; such dust is well known to travel thousands of miles across the oceans. Professor J. M. Prospero, a geophysicist, has regularly found Saharan dust in the air over the West Indies. The Hawaiian islands similarly receive dust from the Asian continent some 4,000 miles distant. The mineral content of this dust when rained out onto the sea may also help the nutrition of the algae there. The surface of the dust particles is not such as to make them suitable as cloud condensation nuclei, but they are washed out of the air by rain induced by the dimethyl sulfide. Lastly, the clouds formed above the ocean filter the radiation reaching the water surface and reduce the proportion of potentially harmful short-wavelength ultraviolet. Visible light needed for photosynthesis is not a limiting factor in the nutrient-poor ecosystems of the oceans, so the shading effect of the clouds is not an adverse one.

None of these effects is large, but taken together they may be enough to improve the meadows of the sea and enable the algal species there to leave more progeny. The geophysiological system requires the continuing production of dimethylsulfonio propionate and of the algae that make it. The difficult question is, How does this system become a part of global climate regulation? The oceans become saltier when water freezes out as ice on the polar surfaces; this might lead to increased emission of dimethyl sulfide, increased cloudiness, and so a positive feedback on further cooling. It might be that the greater biomass associated with the glaciations provides more nutrient for ocean life and so sustains the algae.

As I write, our first scientific paper on this affair has been published in *Nature*. These are the early days of this research, and already it looks like becoming an exciting scientific area for research. Two groups of French glaciologists—Robert Delmas and his colleagues, and C. Saigne and M. Legrand—have recently reported their discovery of sulfuric and methane sulfonic acids in antarctic ice cores, going from the present to 30,000 years ago. Their data shows a strong inverse correlation between global temperature and the deposition in the ice of these acids. Sulfuric acid has several natural sources, but methane sulfonic acid is unequivocally the atmospheric oxidation product of dimethyl sulfide. There was 2 to 5 times larger a deposition of this substance during the ice age and it seems probable that this was due to a greater output from the ocean ecosystems. If confirmed, it suggests that cloud cover and low carbon dioxide operated in synchrony as part of a geophysiological process to keep the Earth cool. More conservative scientists favor a geophysical explanation arising from the theory of the ocean scientist W. S. Broecker, who has proposed that the glaciations are associated with large-scale changes in the circulation of water in the oceans. Certainly the increase in supply of nutrients that would accompany such an event would alter biological productivity and hence the rate of removal of carbon dioxide and the production of dimethyl sulfide. It looks like becoming an interesting debate.

I thought that it might be useful to end this section with a geophysiologist's view of the evolution of the climate and the chemical composition of the atmosphere (see figure 6.3). It is a view of long periods of homeostasis punctuated by large changes.

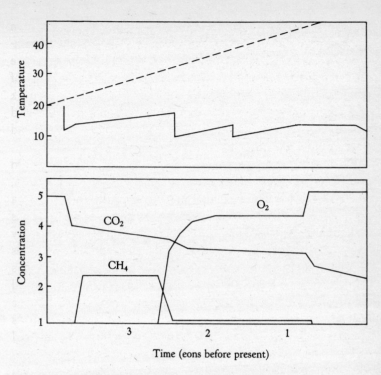

6.3 A geophysiologist's view of the evolution of the climate and atmospheric composition during the life span of Gaia. The upper panel compares the probable temperatures in the absence of life with the stepped but long-term constancy of the actual climate. The lower panel illustrates the stepped fall of carbon dioxide steadily from 10 to 30 percent to its present low level of about 300 parts per million. The early dominance of methane and later of oxygen is also shown. The scale of gaseous abundance is in parts per million by volume and in logarithmic units so that 1 equals 10 parts per million and 5 equals 100,000 parts per million.

We seem to be approaching the end of one of these long stable periods. When life began, the Sun was less luminous and the threat was overcooling. In the middle ages of the Proterozoic, the Sun shone just right for life and little regulation was needed, but now it grows hot and overheating becomes an ever-increasing threat to the biosphere of which we are a part.

7

Gaia and the Contemporary Environment

A day like today I realize what I've told you a hundred different times—that there is nothing wrong with the world. What's wrong is our way of looking at it.

HENRY MILLER, A Devil in Paradise

A rough, stony track led forward across the thin moorland grass and then dropped into the bare, rock-strewn bed of the River Lydd; straight ahead rose the small mountain of Widgery Tor, a sort of turret on the walls of that castle-like massif that was Dartmoor. On a clear, sunny day it was a grand and romantic prospect to reward the small effort of a country walk.

The second day of August 1982 was just such a sunny day, but the moorland vision was all but lost in a dense and dirty brown haze. The air was corrupted by the fumes of Europe's teeming millions of cars and trucks. Their flatus oozed in a gentle easterly drift of wind from the continent; ineluctable chemistry driven by sunlight stewed the fumes into a witches' brew that seared the green leaves. Even my eyes, though washed by the flow of tears, began to smart; soon personal discomfort drew my attention from the contemplation of the mask of photochemical smog that obscured the jeweled brightness of the West

Country scene. A visitor from Los Angeles would instantly have recognized it for what it was, but Europeans, still in the late stages of their honeymoon with personal transport, cannot admit to themselves that their beloved cars fart anything so foul as smog.

This vision of a blighted summer's day somehow encapsulates the conflict between the flabby good intentions of the humanist dream and the awful consequences of its near realization. Let every family be free to drive into the countryside so that they can enjoy its fresh air and scenic beauty; but when they do, it all fades away in the foul haze that their collective motorized presence engenders. As I climbed the Tor and thought these thoughts I knew also that, in driving myself to the foothills at the start of the walk, I had added a small but culpable increment of hydrocarbons and of sulfur and nitrogen oxides to the over-laden air. I also knew that my dislike of this kind of air pollution was a value judgment, and a minority one at that.

There can be very few who do not in some way add to the never-ceasing demolition of the natural environment. Character-istically, arrogantly, we blame technology rather than ourselves. We are guilty, but what is the offence? Many times in the Earth's history new species with some powerful capacity to change the environment have done as much, and more. Those simple bacteria that first used sunlight to make themselves and oxygen were the ancestors of the trees today but eventually, simply by living and by doing their photochemical trick, they so profoundly altered the environment that vast ranges of their fellow species were destroyed by the poisonous oxygen that accumulated in the air. Other simple microorganisms have in their communities acted so that mountain ranges formed and continents were set in motion over the surface of the Earth.

Looked at from the time scale of our own brief lives, environ-mental change must seem haphazard, even malign. From the long Gaian view, the evolution of the environment is character-ized by periods of stasis punctuated by abrupt and sudden change. The environment has never been so uncomfortable as to threaten the extinction of life on Earth, but during those abrupt changes the resident species suffered catastrophe whose scale was such as to make a total nuclear war seem, by comparison, as trivial as a summer breeze is to a hurricane. We are ourselves a product of one such catastrophe. Could it be that we are unwittingly precipitating another punctuation that will alter the environment to suit our successors?

A group of scientists from all parts of the world met in 1984 at São José dos Campos in Brazil. The meeting, held at the request of the United Nations University, posed the question, "How does human intervention in the natural ecosystems of the human tropics affect the forest, the regions around it, and the whole world?" It soon became clear that, whatever their disciplines, the specialists had little to offer other than a frank and honest admission of ignorance. Asked, "When shall we know the consequences of removing the forest from Amazonia?" they could only answer, "Not before the forests have gone." It seemed as if we were at a stage in understanding the health of Gaia rather like that of a physician before scientific medicine existed.

In *The Youngest Science,* Lewis Thomas lets us identify with a young medical practitioner whose first experiences were in the 1930s. Even for those who knew medicine then, it is astonishing to be reminded how little there was that a physician could do to cure a patient. The practice of medicine was largely a matter of administering symptomatic relief and trying to insure that the environment of the patient was that most favorable for the powerful natural processes of healing that we all possess.

Early in its history, medical wisdom accumulated by shrewd observation and by trial and error. The discovery of the curative value of drugs like quinine, or of that wonderful panacea for pain and discomfort, opium, was not in some brightly lit laboratory. Rather, it was in the early experiments or observations of a village genius who realized that there were real benefits to be had from chewing that bitter bark of the cinchona tree or real comfort implicit in the dried latex of the poppy head.

Physiology, the systems science of people and animals, was at first an unrecognized background, but later came to influence further progress. The recognition by Paracelsus that the poison is the dose is a physiological enlightenment still to be discovered by those who seek the unattainable and pointless goal of zero for pollutants. The discovery by William Harvey of the circulation of the blood added to the wisdom of medicine, just as the discovery of meteorology added to our understanding of the Earth. The expert sciences of biochemistry and microbiology came much later, and it was a long time before their new knowledge could enhance the practice of medicine. Even as I write, a paper has appeared in *Nature* describing the molecular structure of the virus responsible for the acquired immune deficiency syndrome; but it will be a long time before this astonishing feat of biochemistry rescues those now dying of AIDS and comforts those who fear that strange and deadly malady.

It seemed oddly appropriate to gather in Brazil; as if we were old-style clinicians conferring at the bedside of a patient with an untreatable disease. We recognized the inadequacy of our expertise and the need for a new profession: planetary medicine, a general practice for the diagnosis and treatment of planetary ailments. We thought that it would grow from experience and empiricism just as medicine had done. It also seemed to some of us that geophysiology, the systems science of the Earth, might serve as did physiology in the evolution of medicine, as a scientific guide for the development of this putative profession.

This chapter, therefore, will be a look at the real and imaginary problems of Gaia through the eyes of a contemporary practitioner of planetary medicine. The scientific background, geophysiology, has already been touched on in the preceding chapters. So let us look at the physical signs, the clinical features, to see if anything can be diagnosed. It is true that, in the case of Gaia, the complaint comes not from the patient but rather from the intelligent fleas that infest her. There is nothing to stop us, however, from going through a routine examination of the temperature charts and the biochemical analyses of the body fluids.

The Carbon Dioxide Fever

Carbon dioxide is a colorless gas with a faintly pungent odor and acid taste. It occurs naturally in the Earth's atmosphere where it serves as an essential plant nutrient and an important determinant of the Earth's thermal balance. Human activities release carbon dioxide to the atmosphere through the burning of wood, coal, oil, natural gas, and other organic materials. Due at least in part to these activities, the concentration of carbon dioxide has increased some 7 percent over the last two decades. There has been much debate over how and when the Earth will respond and what impact this will have on mankind.

So begins the first chapter of that splendid book, *Carbon Dioxide Review 1982,* edited by William Clark. Unless some significant new discoveries have been made between the time of its publication and your reading of this, I suspect that it will still be among the best sources of information on this complex subject.

From the very beginning of life on Earth, carbon dioxide has had a contradictory role. It is the food of photosynthesizers and therefore of all life; the medium through which the energy of sunlight is transformed into living matter. At the same time it has served as the blanket that kept the Earth warm when the Sun was cool, a blanket that, now the Sun is hot, is becoming thin; yet one that must be worn, for it is also our sustenance as food. We have seen earlier how the biota everywhere on the land and sea are acting to pump carbon dioxide from the air so that the carbon dioxide which leaks into the atmosphere from volcanoes does not smother us. Without this never-ceasing pumping, the gas would rise in concentration within a hundred thousand years to levels that would make the Earth a torrid place, and unfit for almost all life here now. Carbon dioxide for Gaia is like salt for us. We cannot live without it, but too much is a poison.

For humans, a hundred thousand years is almost indistinguishable from infinity; to Gaia, who is about 3.6 eons old, it is equivalent to no more than three of our months. Gaia has cause for concern about the long-term decline of carbon dioxide, but the rise of carbon dioxide from burning fossil fuels is, for her, just a minor perturbation that lasts but an instant of time. She is, in any case, tending to offset the decline.

Before we dismiss Gaia from our worries about carbon dioxide, we should bear in mind that among the things that can happen in an instant is the impact of a bullet in full flight. Small it may be, and short the time of contact, but disastrous are its consequences. So it could be with Gaia and carbon dioxide. Humans may have chosen a very inconvenient moment to add carbon dioxide to the air. I believe that the carbon dioxide regulation system is nearing the end of its capacity. The air in recent times has been uncomfortably thin in carbon dioxide for mainstream vegetation and, as was mentioned in the last chapter, new species with a different biochemistry are evolving. These new species, the C4 plants, can live at very low carbon dioxide levels and might at some future time replace the older obsolete C3 models, as the gas continues its progressive down-ward course. The progression is not a smooth one, but more like the trembling and jerky gait of the elderly. We know that carbon dioxide has fallen in abundance during the Earth's history, but it jumped from 180 to near 300 parts per million within a hundred years as the last glaciation ended. A rapid rise like this can have come only from the sudden failure of the pumps. It cannot be explained by the slow processes of geochemistry.

The rate and the extent of the rise of carbon dioxide now under way as a result of our actions is comparable with that of the natural rise that terminated the last ice age. Some time in the next century it seems likely that the increment we add will be equal to that caused by the failure of the pumps some 12,000 years ago. The change of climate we need to think about, therefore, is possibly one as large as that from the last ice age until now; one that would make winter spring, spring summer, and summer always as hot as the hottest summer you can recall. To comprehend such a change at the personal level, imagine you are a citizen of a mid-continental town such as Chicago or Kiev. The change is as great as from the bitter cold of winter that has passed to the fierce heat of summer soon to come.

In his book, William Clark compares the predictions that economists have made of growth from now until the middle of the next century. Among them is listed the prediction by Amory and Hunter Lovins, who argue plausibly that growth may be close to zero into the foreseeable future. This is a very different prediction from that of the late, great Herman Kahn, who saw the whole world in the next century as a vast and wealthy suburban development. Scarsdale writ large. There is strong objective evidence from the record of industrial production that the Lovins' prediction is nearer the truth. Since 1974 the turnover of energy and materials by the human world has been in steady state. Even so, unless we greatly decrease the rate of burning fossil fuel, the atmospheric carbon dioxide will continue to rise to its own steady state and will have doubled in concentration by between 2050 and 2100.

I can only guess the details of the warm spell due. Will Boston, London, Venice, and the Netherlands vanish beneath the sea? Will the Sahara extend to cross the equator? The answers to these questions are likely to come from direct experience. There are no experts able to forecast the future global climate.

Some wisdom comes from geophysiology, which reminds us that the Earth is an active and responsive system and not just a damp and misty sphere of rock. Systems in homeostasis are forgiving about perturbations, and work to keep the comfortable state. Maybe, if left to herself, Gaia could absorb the excess carbon dioxide and the heat that it brings. But Gaia is not left to herself; in addition to carbon dioxide increments, we are also busy removing that part of the plant life, the forests, that by responsive extra growth might serve to counteract the change.

Much more serious than the direct and predictable effects of adding carbon dioxide to a stable system are the consequence of disturbing a system that is precariously balanced at the limits of stability. From control theory, and from physiology, we know that the perturbation of a system that is close to instability can lead to oscillations, chaotic change, or failure. Paradoxically, an animal close to death from exposure to cold, whose core temperature is below 25°C, will die if put into a warm bath. The well-intentioned attempt to restore heat succeeds only in warming the skin to the point where its oxygen consumption is greater than the slowly beating, still-cold heart and lungs can supply. In a vicious circle of positive feedback the blood vessels of the skin dilate; this so reduces blood pressure that death comes rapidly from the failure of the heart as a pump to circulate blood that is too depleted in oxygen for the system's needs. A hypothermic animal will recover if left to warm slowly, or if heat is supplied internally as by diathermy.

We know too little about the carbon dioxide climate system to be able to provide a detailed forecast of the consequences of the current increase, but there are some solid facts of observation from which some general conclusions can be drawn. The Earth's mean temperature is well below the optimum temperature for plant life. There are periodic climatic oscillations as we cycle between the glaciations and their intermissions, and carbon dioxide is attenuated close to its lower possible limit. All these are physical signs of a system on the verge of failure.

Like our latter-day physician, we find that diagnosis is easier than a cure. We are left with the uneasy feeling that to add carbon dioxide to the Earth now could be as unwise as warming the surface of our hypothetical hypothermic patient. It is not much comfort to know that, if we inadvertently precipitate a punctuation, life will go on in a new stable state. It is a near certainty that the new state will be less favorable for humans than the one we enjoy now.

A Case of Acid Indigestion

The greenhouse effect of carbon dioxide is not the only problem to arise from the burning of fossil fuels. In the northern temperate regions of the Earth there is an increased morbidity and mortality of the ecosystems. Trees, and the life in lakes and rivers, are particularly affected. The symptoms seem to be connected with an observed increase in the rate of deposition of acidic substances. Combustion is said to be the cause of acid deposition and of all the harm it does to forest ecosystems. Does geophysiology have any different view on this?

. It could be said that it is all the fault of oxygen. If those ancient godfathers, the cyanobacteria, had not polluted the Earth with this noxious gas there would be no oxides of nitrogen and of sulfur to trouble the air, and therefore no acid rain. Oxygen, the acid maker, the gaseous drug that both gives us life and kills us in the end, not for nothing did those French chemists of the eighteenth century call it the acidifying principle. In their time there were not many chemicals for them to experiment with; those they did have, such as sulfur, carbon, and phosphorus, all gave acids when combined with oxygen. It was only later, when the discovery of electricity allowed chemists to isolate elements like sodium and calcium, that combustion was found to produce alkalis as well. Later still, they realized that an acid was a substance that freely donated positively charged hydrogen atoms, and it was these protons that were the true principle of acidity. In addition that great chemist, G. N. Lewis, observed that it was the electric charge that mattered, not the atom that carried it. He showed that acids can be substances that attract electrons, the fundamental carriers of the negative charge. In some ways oxygen itself is one of these "Lewis" acids.

It is not so surprising that there is free oxygen in the air from life's chemical transactions. The bundle of elements that form the chemicals that go to make up the Earth's crust have more oxygen than anything else; 49 percent of the elemental composition is oxygen. As Lavoisier observed, of all the principal light elements that go to make up living matter—carbon, nitrogen, hydrogen, sulfur, and phosphorus—only hydrogen does not give acids when combined with oxygen. Long before humans trod the planet, the rain that fell was acid. The natural acid in the rain were carbonic acid, the gentle acid of fizzy carbonated water; formic acid, one of the end products of methane oxidation; and nitric, sulfuric, methanesulfonic, and hydrochloric acids. Although the last four of these are strong and corrosive, the rain that fell did no harm, for the acids were present at great dilution. They came mostly from the oxidation of gases emitted by living things; some came also from the gases vented by volca- noes, or from high-energy processes, such as lightning and cosmic rays, that cause nitrogen and oxygen to react. The biological precursors of the acids—for example methane, nitrous oxide, dimethyl sulfide, and methyl chloride—are not acids, but they oxidize in the air to produce the catalog of acids listed above.

Pollution by acid rain deposition is again a matter of dosage: pollution is due to an increase to intolerable levels of acids that previously were benign in their abundance. Quite separate from the demolition of ecosystems by acids and oxidants is the reduction of the quality of life by this kind of pollutant. The smog and haze that I complained of in the opening paragraphs of this chapter, and that masks so much of the Northern Hemi- sphere in summer, is for the most part a fog of sulfuric acid droplets.

Any detached observer of the heated European or North American debate over acid rain might gather the impression that all acid rain was due to the burning of sulfur-rich fossil fuel in power stations, industrial furnaces, and domestic heating systems. Coal and oil both contain about one percent of sulfur. This element leaves the chimney stacks as sulfur dioxide gas, and soon this gas is oxidized to sulfuric acid which condenses as droplets that attract water vapor from the air to form an acid fog or haze. Eventually, this either settles out or is rained out. Where it falls on land that is rich in alkaline rocks like limestone, and particularly if there is a shortage of sulfur there, its fallout is welcome. But when it falls on land that is already acid, its addition is unwelcome and potentially destructive. Can- ada, Scandinavia, Scotland, and many other northern regions

are on ancient rocks, the hard, soluble residue of eons of weathering. The ecosystems that survive on this unpromising, and often
normally acidic, terrain have less capacity to resist the stress
of acidification. It is from the countries of these regions that
comes a justifiable complaint that their industrial neighbors are
destroying them. To the Canadians and the Scandinavians it
seems unarguable that the emission of sulfur dioxide by countries
downwind should cease. Few can doubt the natural justice of
their case, but naturally the offenders are reluctant to spend
the very large sums needed to stop the escape of sulfur dioxide
from their power stations and industries.

The geophysiological contribution to this debate is to observe
that this acid indigestion may have another source in addition
to the sulfuric vinegar of neighbors. The fitting of sulfur dioxide
removers to the chimneys might only alleviate, not cure, the
problem. The neglected source of acid is the natural sulfur carrier,
dimethyl sulfide. In the past two years, Meinrat Andreae and
Peter Liss (ocean chemists based, respectively, in Florida and
the United Kingdom) have shown that the emission of this
gas from phytoplankton blooms at the surface of the oceans
around western Europe is so large as to be comparable with
the total emissions of sulfur from industry in this region. Moreover, the phytoplankton emissions are seasonal and seem to
coincide with the maximum of acid deposition.

It might be asked, with reason, that if this is the case, why
was the pollution not observed until recently? If dimethyl sulfide
from the sea is the source of sulfuric acid, then surely wouldn't
Scandinavia always have suffered the ill effects of acid deposition? In fact, two changes in recent years may have made the
natural transfer of sulfur from the sea to the land a curse instead
of a benefit. Before Europe became intensively industrialized,
dimethyl sulfide from the sea was probably carried far inland
by the westerly wind drift, and slowly dropped its sulfur in
dilute form over a vast area. Industrialization has not only increased the total burden of acid but also has greatly increased
the abundance of oxides of nitrogen and other chemicals from
combustion. In sunlight, these can react to make the powerful

oxidant hydroxyl. Most important as a source of these agents is the internal combustion engine that powers personal and public transport. Hydroxyl radicals are now locally at least ten times more abundant than they used to be before private transport became ubiquitous. Because of this, dimethyl sulfide that used to oxidize slowly over all of Europe may now dump its burden of acid rapidly over the regions near the coast where the sea air encounters the polluted air.

In addition to this increase in the rate of oxidation, and hence acid production, the output of dimethyl sulfide has itself probably increased in recent years. Patrick Holligan, of the Marine Biological Laboratory in Plymouth, tells me that satellite photographs have revealed dense algal blooms clustered around the outlets of the continental rivers of Europe. Peter Liss and his colleagues have found that these algal blooms emit dimethyl sulfide, apparently stimulated by the rich flow of nutrients down the rivers of Europe. The excessive use of nitrate fertilizer, and the increased output of sewage effluent into the rivers feeding the North Sea and the English Channel, have gone to overnourish the sea above the European continental shelf and to make it like a duck pond. The relative amount of acid from this source is not yet known. It might turn out to be insignificant. However, prudent legislators concerned over acid rain should urge their scientific advisers to investigate the relative importance of the various sources of acid. My personal sympathy is with those who ask for action on the basis that sulfur dioxide emissions are the prime culprit. I do wonder, though, what would happen if reducing these did not work. Would governments then have the will to tackle the very much more expensive acts, if these were the best way to prevent acid rain deposition, of sewage reform, or the control of nitrogen oxide emissions?

The affair of acid rain is as much an issue of politics and economics as of environmental science. Before accepting as inevitable a long and costly battle involving national and commercial interests, it is useful to go back and re-examine the conduct of the ozone war. There are some interesting parallels and possibly some lessons to be learnt.

The Dermatologists' Dilemma: Ozonemia

In the late 1960s I developed a simple apparatus able to detect chlorofluorocarbons (CFCs) in the atmosphere down to parts per trillion by volume. This is an exquisite sensitivity; at such levels even the most toxic of chemicals could be breathed in or swallowed without harm, indefinitely. In 1972, I took this apparatus on the voyage to Antarctica and back of the RV *Shackleton* (see chapter 6). The measurements I made on that ship showed that the CFCs were distributed throughout the global environment. There was about 40 parts per trillion in the Southern Hemisphere and 50 to 70 in the Northern. When I reported these findings in a *Nature* paper in 1973, I was concerned that some enthusiast would use them as the basis of a doom story. As soon as numbers are attached to the presence of a substance, these numbers seem to confer a spurious significance. What previously was a mere trace becomes a potential hazard. The hypochondriac, on hearing that his blood pressure is 110/60, becomes worried: "Surely, doctor, isn't that too low?" As a putative planetary physician I felt the need to add at the beginning of my paper the sentence, "The presence of these compounds constitutes no conceivable hazard." This sentence has turned out to be one of my greatest blunders. Of course I should have said, "At their present level, these compounds constitute no conceivable hazard." Even then I knew that, if their emissions continued unchecked, they would accumulate until sometime near the end of the century they could be a hazard. I knew nothing of their threat to the ozone layer, but I did know that they were among the most potent of greenhouse gases and that by the time they reached the parts per billion level the climatic consequences of their presence could be serious. This opinion is recorded in the proceedings of a conference on fluorocarbons held at Andover, Massachusetts, in October 1972.

At this time in the 1970s there was a fear of impending catastrophe. "Earth's fragile shield," the ozone layer, was said to be in imminent danger of demolition as a consequence of the release into the stratosphere of nitric oxide in the exhaust gases of supersonic aircraft. The atmospheric chemist Harold Johnson first alerted us to this particular threat. Then, tentatively at first, Ralph Cicerone and his colleague Richard Stolarski drew our attention to chlorine as an another danger to ozone. Then in 1974 there appeared in *Nature* a paper by Sherry Rowland and Mario Molina which argued with great clarity and force that the CFCs, as a result of stratospheric photochemistry, were a potent source of chlorine and hence a threat to ozone. This paper stands like a beacon, a natural successor to Rachel Carson's book *Silent Spring*. It heralded the start of the ozone war. In their enthusiasm with the science and the battle, scientists, somewhat uncharacteristically, convinced themselves and the public of the need for immediate action to ban the emission of CFCs. To me, wrong-footed by my earlier assertion that CFCs were harmless, it seemed to be a remote and hypothetical threat. But I was among a minority, and legislators in many parts of the world were persuaded to act precipitately and to enact legislation banning CFC gases as aerosol propellants. It is interesting to ask what is special about ozone that made legislators act this way. No one was dying of the effects of CFC emissions; crops and livestock were unharmed by their presence; the substances themselves were among the most benign of chemicals that enter our homes, neither toxic, corrosive, nor flammable. Indeed, they would have been imperceptible but for the sensitivity of the instrument that I used for their detection. Their presence at between 40 and 80 parts per trillion, even to the most committed environmentalist, was no threat to ozone. The concern came from the fact that the emissions were growing exponentially, and if the growth rate of the sixties continued until the end of the century, there would be an ozone depletion of between 20 and 30 percent. This would be disastrous.

Ozone is a deep blue, explosive, and very poisonous gas. It is strange that so many have regarded it as if it were some beautiful endangered species. But it was the mood of the 1970s to respond to environmental hazards much as previous generations had responded to witchcraft. It was not easy to oppose the widely held belief that only immediate action by scientists and politicians could save us and our children from an otherwise ineluctable depletion of the ozone layer and the dire consequences of an ever-increasing flux of carcinogenic ultraviolet radiation. This was also the time when the word "chemical" became pejorative, and all products of the chemical industry were assumed to be bad unless proved harmless. In a more sensible environment, we might have regarded the predictions of doom in the next century due to a single industrial chemical as far fetched—something to watch closely, but not something requiring immediate legislation. But the 1970s was not the time for a long, cool look at things.

At a university in the small Rocky Mountain town of Logan, Utah, the principal scientists and lawyers concerned with the fluorocarbon affair met in 1976. Among those present were Ralph Cicerone, who had first hypothesized that chlorine in the stratosphere might catalyze the destruction of ozone, and Mario Molina and Sherry Rowland, who had developed the complex reaction sequence that explained how the CFCs could be the source of the chlorine and delineated the intricate details of the destruction mechanism. There were also scientists from industry and from the regulatory agencies, and there were, of course, lawyers and legislators. It could have been a reasoned debate leading to agreement about a safe upper limit for the CFCs in the light of current knowledge. It was instead a kind of tribal war council where the decision to fight was taken. Anyone who was not for the immediate banning of the CFCs was clearly a traitor to the cause. I shall never forget the adversarial encounter between Commissioner R. D. Pittle and Dr. Fred Kaufman, who was representing the National Academy of Sciences. The commissioner forgot he was not in a courtroom and demanded a yes or no answer to the question of whether CFCs should be banned. In certain ways it reminded me of another encounter, long ago: the one between Galileo and the authorities of his time.

The processes of science are very different from those of the courtroom. Both have evolved to satisfy the needs of their practioners. Scientific hypotheses are best tested by the accuracy of their predictions; the establishment of a fact of science does not greatly affect the Universe, only the wisdom of scientists. By contrast, facts in law are tested in an adversarial debate and established by judgment. The establishment of a legal fact alters society from then on. At the best of times, and even with near certainties, science and the law do not mix well. At Logan they tried to form legal judgments on plausible but untested scientific hypotheses. It is not so surprising that the result was of little credit to any of the participants.

Once again the wisdom of Paracelsus that the poison is the dose was ignored, and in its place the "zero" shibboleth took charge. "There is no safe level of ultraviolet radiation," was the cry. "Ultraviolet, like other carcinogens, should be reduced to zero." In fact, ultraviolet radiation is part of our natural environment, and has been there as long as life itself. It is the nature of living things to be opportunistic. Ultraviolet, although potentially harmful, can also be used by living organisms for the photosynthesis of vitamin D. When it is a threat, it can be avoided by synthesizing such pigments as melanin to absorb it.

There is still a lack of knowledge about the relationships between natural ecosystems and the ultraviolet to which they are exposed. But we do know that ultraviolet radiation varies sevenfold (700 percent) in intensity between the Arctic and the tropics, whereas visible light varies only 1.6 times (160 percent) over the same range of latitude. In spite of this large range of intensity, there is nowhere a region where the growth of vegetation is limited by ultraviolet. In contrast, a sevenfold change of rainfall makes the difference between forests and deserts. There are no ultraviolet deserts on Earth, and life seems well adapted to the radiation over this wide range of intensity. Damage does occur but seems to be limited to recent migrants from high to low latitudes. There is also evidence that a lack of ultraviolet can be harmful to migrants from the tropics to the temperate regions.

Exposure to any radiation with a high quantum energy that

penetrates the skin can damage the genetic material of our cells and corrupt their program of instructions. Among the adverse effects is the conversion from normal to malignant growth. This is frightening stuff, but we can keep our cool by remembering that these carcinogenic consequences are no different from those of breathing oxygen, which is also a carcinogen. Breathing oxygen may be what sets a limit to the life span of most animals, but not breathing it is even more rapidly lethal. There is a right level of oxygen, namely 21 percent; more or less than this can be harmful. To set a level of zero for oxygen in the interests of preventing cancer would be most unwise.

Wars do not usually start from a single isolated incident, and so it was with the ozone war. The historical basis was, as mentioned in chapter 4, the proposal by Berkner and Marshall that the colonization of the land surfaces of the Earth did not take place until oxygen and its allotrope, ozone, first entered the atmosphere. Ozone, they said, prevents the penetration of hard ultraviolet radiation that otherwise would keep the land sterilized and uninhabitable by life. This was a decent scientific hypothesis and a very testable one at that. Indeed it was tested by my colleague Lynn Margulis, who challenged it by showing that photosynthetic algae could survive exposure to ultraviolet radiation equivalent in intensity to that of sunlight unfiltered by the atmosphere. But this did not stop the hypothesis from becoming one of the truly great scientific myths of the century; it is almost certainly untrue, and it survives only because of the apartheid that separates the sciences. Physical scientists regard biology as extraterritorial and biologists reciprocate. The members of each discipline tend to accept uncritically the conclusions of the other. This apartheid is a triumph of expertise over science, and it is expressed with great innocence when scientists try to explain the separation of their findings into physical and biological parts as a necessary consequence of expertise. Biologists concerned with the effects of ultraviolet know it to be beneficial as well as harmful. But until recently they had no reason to doubt the expertise of their physical science colleagues, and therefore thought only of the consequences of ozone depletion. As a counterpoint, most physical scientists are unaware that ultraviolet might in any way have benefits. Consequently, they tend to think of ozone accretion as a benefice. However, the diseases of vitamin D deficiency—rickets and osteomalacia—are associated with a reduced exposure to solar ultraviolet. Also it seems that the incidence of multiple sclerosis varies with latitude reciprocally to that of skin cancer. The variation of skin color with latitude suggests that we have, in the absence of migration, adapted to the ultraviolet levels of our habitats.

Once more ozone is news. J. C. Farman and B. G. Gardiner, of the British Antarctic Survey, have discovered a thinning of the ozone layer over the south polar regions, moreover a thinning that has grown rapidly each year until now it is almost a hole. This event is entirely unexpected and in great contrast to the fact that over most of the world the level of ozone is either unchanged or even slightly increased. But this is exciting and fearful stuff. What if the hole should spread and threaten popu-lated regions? Before we become too deeply involved, it seems worth asking what were the benefits of the first ozone conflict? Who won and who lost? The only clear losers were those small industries, and their employees, dependent upon the use of CFCs in the products that were banned. For various and complex reasons, the manufacturers of CFCs were not much affected. The loss of the doubtfully profitable CFC-propellant section of their market, together with the rationalization of their industry, did little to change their economy. Politicians and the environ-mental movement lost some of their credibility, but public mem-ory tends to be short. The clear winners were science and the scientists. Vast sums have been disbursed for atmospheric re-search, which would never have been available but for the ozone war. We now know much more about our atmosphere, and this knowledge will be essential in the understanding of other atmospheric problems. Among them is the greenhouse warming effect of minor atmospheric gases. Three key properties of the CFCs make them dangerous. First is their long atmospheric life times, which allow them to accumulate unchecked, second their ability to carry their burden of chlorine directly and without loss to the stratosphere, and third the intensity with which they absorb long-wavelength infrared radiation. Their presence in the atmosphere adds to the carbon dioxide greenhouse effect. This is a danger that is potentially more serious than that of ozone depletion. We have reason to be glad that one of the pioneers of the original concern about CFCs, Ralph Cicerone, has turned his attention to the graver and more certain dangers of their greenhouse effect.

It may turn out that I was very wrong to have opposed those who sought instant legislation to stop the emission of CFCs. I regard the strange phenomenon over the south polar regions as a warning of other more serious surprises yet to come. It seems possible that other changes, including the concomitant increase of CO_2 and methane from human industry and agriculture, are responsible for the extra effect of chlorine compounds in polar regions, but there is little doubt in my mind that without the chlorine from industrial gases there would be no thinning of the ozone layer at the South Pole. The CFCs and other industrial halocarbons have increased by 500 percent since I first measured them in 1971. They were harmless then, but now there is too much halocarbon gas in the air. The first symptoms of poisoning are now felt. I now join with those who would regulate the emissions of the CFCs and other carriers of chlorine to the stratosphere.

To return to our clinical analogy, we could say that the fear of skin cancer as a consequence of ozone depletion led at first to a global hypochondria—something all too easily acquired by identifying our fears with the plausible account of symptoms described in a textbook. Good physicians know that hypochondria can be a cry for help and mask the existence of a real malady; perhaps the same is true of the global state of health. Could fears about the CFCs and the ozone layer have presaged discovery of the ozone hole and the climate-threatening greenhouse effect of CFCs?

A Dose of Nuclear Radiation

Carl Sagan once observed that if an alien astronomer were to look at the Solar System in the radio-frequency part of the spectrum, it would see a truly remarkable object. Two stars eclipsing one another: one of them a normal, small, main-sequence star and the other a very small, but intensely luminous body with an apparent surface temperature of millions of degrees, our Earth. Were that distant observer a scientist, it might speculate on the nature of the energy source that powered, what seemed to be, one of the hottest objects in the Galaxy. I wonder how high on the list of probable sources it would place chemical energy. Would it include energy coming from the reaction between fossil fuels and oxygen from plants?

It is easy to ignore the fact that we are the anomalous ones. The natural energy of the Universe, the power that lights the stars in the sky, is nuclear. Chemical energy, wind, and water wheels: such sources of energy are, from the viewpoint of a manager of the Universe, almost as rare as a coal-burning star. If this is so, and if God's Universe is nuclear-powered, why then are so many of us prepared to march in protest against its use to provide us with electricity?

Fear feeds on ignorance, and a great niche was opened for fear when science became incomprehensible to those who were not its practitioners. When X rays and nuclear phenomena were discovered at the end of the last century, they were seen as great benefits to medicine—the near-magic sight of the living skeleton and the first means to palliate, even sometimes cure, cancer. Roentgen, Becquerel, and the Curies are remembered with affection for the good their discoveries did. Sure enough, there was a dark side also, and too much radiation is a slow and nasty poison. But even water can kill if too much is taken.

It is usually assumed that the change in attitude towards radiation came from our revulsion at that first misuse of nuclear energy at Hiroshima and Nagasaki. But it is not that simple. I well remember how the first nuclear power stations were a source of national pride as they quietly delivered their benefice of energy without the vast pollution of the coal burners they replaced. There was a long spell of innocence between the end of the Second World War and the start of the protest movements of the 1960s. So what went wrong?

Nothing really went wrong, it just happens that nuclear radiation, pesticides, and ozone depleters share in common the property that they are easily measured and monitored. The attachment of a number to anything or anyone bestows a significance that previously was missing. Sometimes, as with a telephone number, it is real and valuable. But some observations—for example, that the atmospheric abundance of perfluoromethyl cyclohexane is 5.6×10^{-15}, or that as you read this line of text at least one hundred thousand of the atoms within you will have disintegrated—while scientifically interesting, neither confer benefit nor have significance for your health. They are of no concern to the public.

But once numbers are attached to an environmental property the means will soon be found to justify their recording, and before long a data bank of information about the distribution of substance X or radiation Y will exist. It is a small step to compare the contents of different data banks and, in the nature of statistical distributions, there will be a correlation between the distribution of substance X and the incidence of the disease Z. It is no exaggeration to observe that once some curious investigator pries open such a niche, it will be filled by the opportunistic growth of hungry professionals and their predators. A new subset of society will be occupied in the business of monitoring substance X and disease Z, to say nothing of those who make the instruments to do it. Then there will be the lawyers who make the legislation for the bureaucrats to administer, and so on. Consider the size and intricacy of the radiation-monitoring agencies, of the industry that builds monitoring and protective devices, and of the academic community that has radiation biology as its subject. If the strong public fear of radiation were dispelled, it would not be helpful to their continued employment. We see that there is a very biological, Gaian, feedback in our community relationship with the environment. It is not a conspiracy or a selfishly motivated activity. Nothing like that is needed to maintain the ceaseless curiosity of explorers and investigators, and there are always opportunists waiting to feed on their discoveries.

If this alone were not enough, there are the media, ready to entertain us. They have in the nuclear industry a permanent soap opera that costs them nothing. Why, we can even experience the excitement of a real disaster, like Chernobyl, but in which, as in fiction, only a few heros died. It is true that calculations have been made of the cancer deaths across Europe that might come from Chernobyl, but if we were consistent, we might wonder also about the cancer deaths from breathing the coal smoke smogs of London and look on a piece of coal with the same fear now reserved for uranium. How different is the fear of death from nuclear accidents from the commonplace and boring death toll of the roads, of cigarette smoking, or of mining—which when taken together are equivalent to thousands of Chernobyls a day.

It was Rachel Carson, with her timely and seminal book, *Silent Spring,* who started the Green Movement and made us aware of the damage we can so easily do to the world around us. But I do not think that she could have made her case against pesticide poisoning without the prior discovery that agricultural pesticides were distributed ubiquitously throughout the whole biosphere. Numbers could even be attached to the wholly insignificant quantities of pesticide in the milk of nursing mothers or in the fat of penguins in the Antarctic. In Rachel Carson's time, pesticides were a real threat, and the blind exponential increase of their use put all our futures in hazard. But we have responded in a fashion, and that one experience ought not be extrapolated to all environmental hazards real or imagined.

The foregoing paragraphs are not intended as support for the nuclear industry, nor to imply that I am enamoured of nuclear power. My concern is that the hype about it, both for and against, diverts us from the real and serious problem of living in harmony with ourselves and the rest of the biota. I am far from being an uncritical supporter of nuclear power. I often have a nightmare vision of the invention of a simple, lightweight nuclear fusion power source. It would be a small box, about the size of a telephone directory, with four ordinary electricity outlets embedded in its surface. The box would breathe in air and extract, from its content of moisture, hydrogen that would fuel a miniature nuclear fusion power source rated to supply a maximum of 100 kilowatts. It would be cheap, reliable, manufactured in Japan, and available everywhere. It would be the perfect, clean, safe power source; no nuclear waste nor radiation would escape from it, and it could never fail dangerously.

Life could be transformed. Free power for domestic use; no one need ever again be cold in winter or overheated in the summer. Simple, elegant pollution-free private transport would be available to everyone. We could colonize the planets and maybe even move on to explore the star systems of our Galaxy. That is how it might be sold, but the reality almost certainly is ominously expressed by Lord Acton's famous dictum, "Power tends to corrupt and absolute power corrupts absolutely." He was thinking of political power, but it could be just as true of electricity. Already we are displacing the habitats of our partners in Gaia with agricultural monocultures powered by cheap fossil fuel. We do it faster than we can think about the consequences. Just imagine what could happen with unlimited free power.

If we cannot disinvent nuclear power, I hope that it stays as it is. The power sources are vast and slow to be built, and the low cost of the power itself is offset by the size of the capital investment required. Public fears, unreasoning though they sometimes are, act as an effective negative feedback on unbridled growth. No one, thank God, can invent a chain saw driven by a nuclear fission power source that could cut a forest as fast and heedlessly as now we cut down a tree.

To my ecologist friends, many of whom have been at the sharp end of protest against nuclear power, these views must seem like a betrayal. In fact, I have never regarded nuclear radiation or nuclear power as anything other than a normal and inevitable part of the environment. Our prokaryotic fore-bears evolved on a planet-sized lump of fallout from a star-sized nuclear explosion, a supernova that synthesized the elements that go to make our planet and ourselves. That we are not the first species to experiment with nuclear reactors has been touched on earlier in this book.

I am indebted to Dr. Thomas of Oak Ridge Associated Univer-sities, who gave me a new insight on the nature of the biological consequences of nuclear radiation. As I listened to his words, spoken in the quiet privacy of his room, I felt an emotion like that described by Keats in his verses about first reading Chap-man's *Homer*. What Dr. Thomas said may have been no more than hypothetical, but to me it was exciting stuff. Let's look at his proposition: "Suppose that the biological effects of exposure to nuclear radiation are no different from those of breathing oxygen."

We have long known that the agents within the living cell that do damage after the passage of an X-ray photon, or a fast-moving atomic fragment, are an assortment of broken chemicals; things called free radicals that are reactive and destructive chemi-cals. As an X-ray photon passes through the cell, the radiation severs chemical bonds just as a bullet might sever blood vessels and nerves. By far the greater part of this destruction is of molecules of water, for they are the most abundant in living matter. The broken pieces of a water molecule form, in the presence of oxygen, a suite of destructive products including the hydrogen and hydroxyl radicals, the superoxide ion, and hydrogen peroxide. These are all capable of damaging, irrevers-ibly, the genetic polymers that are the instructions of the cell. This is now conventional scientific wisdom; the novel insight from Dr. Thomas was to remind us that these same destructive chemicals are being made all the time, in the absence of radiation, by small inefficiencies in the normal process of oxidative metabo-lism. In other words, so far as our cells are concerned, damage by nuclear radiation and damage by breathing oxygen are almost indistinguishable.

The special value of this hypothesis is that it suggests a rule of thumb for comparing these two damaging properties of the environment. If Dr. Thomas were right, then the damage done by breathing is equivalent to a whole body radiation dose of approximately 100 roentgens per year. I used to wonder about the risk-benefit ratio of a medical X-ray examination. A typical hospital X ray of the chest or abdomen could deliver 0.1 roentgen of radiation, enough to blacken the film of a personal radiation monitor and to have caused terror to the inhabitants of Three Mile Island. Now, thanks to Dr. Thomas, I look upon it as no more than one-thousandth of the effect of breathing for a year. Or to put it another way, breathing is fifty times more dangerous than the sum total of radiation we normally receive from all sources.

The early battles at the end of the Archean against the planet-wide pollution by oxygen are still apparently with us. Living systems have invented ingenious countermeasures: antioxidants such as vitamin E to remove the hydroxyl radicals, superoxide dismutase to destroy the superoxide ion, catalase to inactivate hydrogen peroxide, and numerous other means to lessen the destructive effects of breathing. Nevertheless, it seems likely that the life span of most animals is set by a fixed upper limit of the quantity of oxygen that their cells can use before suffering irreversible damage. Small animals such as mice have a specific rate of metabolism much greater than we do; that is why they live only a year or so even if protected from predation and disease. Oxygen kills just as nuclear radiation does, by destroying the instructions within our cells about reproduction and repair. Oxygen is thus a mutagen and a carcinogen, and breathing it sets the limit of our life span. But oxygen also opened to life a vast range of opportunities that were denied to the lowly anoxic world. To mention just one of these: free molecular oxygen is needed for the biosynthesis of those special structure-building amino acids, hydroxylysine and hydroxyproline. From these are made the structural components that made possible the trees and animals.

Paul Crutzen, an atmospheric chemist, was the first to draw our attention to the far-reaching geophysiological consequences of a major nuclear war, the "nuclear winter." We need to be reminded, often, just how bad that ultimate sanction can be so that it remains a deterrent. But, like oxygen, nuclear energy provides opportunities and challenges us to learn to live with it.

The Real Malady

When things are bad, or if we witness some particularly depressing piece of environmental demolition, we often say that people are like a cancer on the planet; they grow in numbers unchecked and they destroy all that comes in contact with them. Was it fear of cancer, that great standby of all environmentalist demagogues, that stirred our worries about the Earth? If it was, we can cease worrying on that account. Life exists in many forms, and of these, neither organisms living as single cells nor Gaia suffer the unique rebellion of cancer. That is limited to the metaphyta and metazoa—those life forms, such as trees and horses, that consist of vast but intensely organized cell communities. People are not in any way like a tumor. Malignant growth in an animal requires the transformation of instructions encoded in the genes of a cell. The descendants of the transformed cell then grow independently of the animal system. The independence is never complete; the cancerous cells still, to some extent, respond and contribute to the system. To be like a cancer we should need first to become a different species and then to be a part of something far more intensively organized than Gaia.

The longevity and strength of Gaia comes from the informality of the association of her constituent ecosystems and species. She operated for nearly a third of her life populated with no more than prokaryotic bacteria. She still is largely run by these, the most primitive part of life on Earth. The consequences for Gaia of the environmental changes that we have made are as nothing compared with those that you or I would experience from unfettered growth of a community of malignant cells. Although Gaia may be immune to the eccentricities of some wayward species like us or the oxygen bearers, this does not mean that we as a species are also protected from the consequences of our collective folly.

When I wrote the first Gaia book, nearly ten years ago, it seemed that there might be critical ecosystems whose damage or removal might have serious consequences for the present collection of organisms which inhabit the Earth and find it comfortable. The forests of the humid tropics and the ecosystems of the waters of the continental shelves seemed at that time to be those most likely to be crucial for keeping the environmental status quo. Already we see the beginnings of malfunction, in the form of rain that is too acid as a consequence of the prolifera‑ tion of algae in the overnourished waters off the European coast. Also, the general decline of the ecosystems in several parts of Africa may be a consequence of removing the trees that once grew there.

The maladies of Gaia do not last long in terms of her life span. Anything that makes the world uncomfortable to live in tends to induce the evolution of those species that can achieve a new and more comfortable environment. It follows that, if the world is made unfit by what we do, there is the probability of a change in regime to one that will be better for life but not necessarily better for us. In the past, changes of this kind, like the jump from a glaciation to an interglacial, have tended to be revolutionary punctuations rather than gradual evolutions.

The things we do to the planet are not offensive nor do they pose a geophysiological threat, unless we do them on a large enough scale. If there were only 500 million people on Earth, almost nothing that we are now doing to the environment would perturb Gaia. Unfortunately for our freedom of action, we are moving towards eight billion people with more than ten billion sheep and cattle, and six billion poultry. We use much of the productive soil to grow a very limited range of crop plants, and process far too much of this food inefficiently through cattle. Moreover, our capacity to modify the environ‑ ment is greatly increased by the use of fertilizers, ecocidal chemi‑ cals, and earth‑moving and tree‑cutting machinery. When all this is taken into account we are indeed in danger of changing the Earth away from the comfortable state it was once in. It is not just a matter of population; dense population in the northern temperate regions may be less a perturbation than in the humid tropics.

There is no way for us to survive without agriculture, but there seems to be a vast difference between good and bad farming. Bad farming is probably the greatest threat to Gaia's health. We use close to 75 percent of the fertile land of the temperate and tropical regions for agriculture. To my mind this is the largest and most irreversible geophysiological change that we have made. Could we use this land to feed us and yet sustain its climatic and chemical geophysiological roles? Could trees provide us with our needs and still serve to keep the tropics wet with rain? Could our crops serve to pump carbon dioxide as well as the natural ecosystems they replace? It should be possible, but not without a drastic change of heart and habits. I wonder if our great-grandchildren will be vegetarian and if cattle will live only in zoos and in tame life parks.

As understanding about the dangers inherent in farming grows, it reinforces the insight from conventional modeling. Thus large-scale changes of land use, even in one region alone, will not be limited in their effects to that region only. Geophysiology also reminds us that the climatic effects of forest clearance are likely to be additive to those of carbon dioxide and other greenhouse gases. Even the most intricate climate models of the present type cannot predict the consequences of these changes. A complete model requires the biota to be included in a way that recognizes its very active presence and its preference for a narrow range of environmental variables. Putting the biota in a box with inputs and outputs, as in a biogeochemical model, does not do this. By analogy, we need physiology to understand how we sustain a constant personal temperature when exposed to heat or cold, biochemistry can only tell us what reactions produce heat in our bodies, not how we regulate our temperature.

There is as yet no answer as to what proportion of the land of a region can be developed as open farmland or forest without significantly perturbing either the local or the global environment. It is like asking what proportion of the skin can be burnt without causing death. This physiological question has been answered by the direct observations of the consequences of accidental burns. It has not been modeled, so far as I am aware. It may be that detailed geophysiological modeling can answer the parallel environmental question, but, if human physiology is a guide, empirical conclusions drawn from a close study of the local climatic consequences of regional changes of land use are more likely to yield the information we seek.

In some ways the ecosystem of, for example, a forest in the humid tropics is like a human colony in Antarctica or on the Moon. It is only self-supporting to a limited extent, and its continued existence depends upon the transport of nutrients and other essential ingredients from other parts of the world. At the same time, ecosystems and colonies try to minimize their losses by conserving water, heat, or essential nutrients; to this extent they are self-regulating. The tropical rain forest is well known to keep wet by modifying its environment so as to favor rainfall. Traditional ecology has tended to consider ecosystems in isolation. Geophysiology reminds us that all ecosystems are interconnected. As an analogy, an animal's liver has some capacity to regulate its internal environment, and the cells of the liver can be grown in isolation. But neither the animal without a liver, nor the liver itself, can live independently alone; each depends upon the interconnection between the two. We do not know if there are vital ecosystems on the Earth, although it would be difficult to imagine life continuing without the ubiquitous presence of those ancient bacteria who live in the dark and smelly places of mud and feces. Those bacteria found, 3.5 eons ago, that the perfect way of life for them was turning used carbon into methane gas, and they have done it ever since. The ecosystems of the waters of the continental shelves transfer essential elements like sulfur and iodine from the sea to the air and hence to the land. The forests of the humid tropics act on a global scale by pumping vast volumes of water back into the air (evapotranspiration); this has the potential to affect climate locally by causing the condensation of clouds. The white tops of the clouds reflect away the sunlight that otherwise would heat and dry the region. The evaporation of water from the liquid state absorbs a great deal of heat, and the climate of distant regions outside the tropics is considerably warmed when damp tropical air masses release their latent heat in the condensation of rain. The transfer of nutrients and the products of weathering by the tropical rivers are obviously part of their interconnection and must also have a global significance.

If evapotranspiration, or the additions of the tropical rivers to the oceans, is vital to the maintenance of the present planetary homeostasis, then this suggests that its replacement with an agricultural surrogate or a desert not only would deny those regions to their surviving inhabitants but would threaten the rest of the system as well. We do not yet know; we can only guess that tropical forest systems are vital for the world ecology. It may be that they are like the temperate forests that seem to be expendable without serious harm to the system as a whole; temperate forests have suffered extensive destruction during glaciations as well as during the recent expansion of agriculture. It would seem, therefore, that the traditional ecological approach of examining the forest ecosystem in isolation is as important to our understanding as is the consideration of its interdependence with the whole system. Geophysiology is at the information-gathering stage, rather as was biology when Victorian scientists went forth to distant jungles to collect specimens.

We do recognize the needs of the Earth, even if our response time is slow. We can be altruistic and selfish simultaneously in a kind of unconscious enlightened self-interest. We most certainly are not a cancer of the Earth, nor is the Earth some mechanical contraption needing the services of a mechanic.

If it turns out that Gaia theory provides a fair description of the Earth's operating system, then most assuredly we have been visiting the wrong specialists for the diagnosis and cure of our global ills. These are the questions that must be answered: How stable is the present system? What will perturb it? Can the effects of perturbation be reversed? Without the natural ecosystems in their present form, can the world maintain its present climate and composition? These are all within the province of geophysiology. We need a general practitioner of planetary medicine. Is there a doctor out there?

8

The Second Home

Better to be kind at home than to burn incense in a far place.
Chinese Proverb

In the summer of 1969 I was in our second home on the shores of Bantry Bay, that part of Ireland where long thin rocky peninsulas point southwest, like fingers on a hand stretched out towards America. It was Monday, July 21, the day after the astronauts Neil Armstrong and Edwin Aldrin had walked on the Moon. The news of their historic journey came to us by radio. So remote and mountainous was this part of Ireland in those days that we were denied the pleasure of seeing the landing on a television screen. To our family, raised in the contemporary scientific culture, the ascent to the Moon was a consummation. For our Irish neighbors on the Beara Peninsula, it was a mind-quake that shook the foundations of their belief. Throughout the week that followed they often asked us, "Is it really true that men have landed on the Moon?" We were puzzled by the question, and replied, "Of course it is true. Did you not hear it on the radio?" Yes, they had heard it, but they wanted to hear from ourselves that there were men up there on the Moon.

It took a long time and some prompting from my friends and neighbors, Michael and Theresa O'Sullivan, before I realized that what was an undoubted fact for me was, for the different culture that surrounded me, news of a much more profound and deep significance. To many of those living on the remote Beara Peninsula, Heaven was still simply up there in the sky and Hell beneath their feet. Their faith was not perturbed by the news of the men walking on the Moon, but their religious belief seemed to be undergoing an internal reorganization. I can only compare the intensity of their experience with that of the change of mind that came to many in the last century from the news that Darwin brought back after the voyage of the *Beagle*.

In this century it is the tales of astronauts and the harvest of space exploration that has moved the locked plates of our minds. It should not therefore be necessary to explain why there is a chapter about Mars in a book on Gaia, but I will remind you that the Gaia hypothesis was a serendipitous discovery, arising directly from the invention of a method of planetary-life detection intended for use on Mars. Nearly twenty years later I found myself speculating on the possibility of changing the physical environment of Mars so that it becomes a self-sustaining living system and a brother to Gaia. Like the Gaia hypothesis, this notion also had an indirect and unexpected origin, and I shall digress in the next few paragraphs to explain it.

It came about because of a book called *The Greening of Mars*, written with my friend Michael Allaby, a fluent writer on environmental topics. He wanted a world on which to act out a new colonial expansion; a place with new environmental challenges and free of the tribal problems of the Earth. I just wanted a model planet on which to play new games with Gaia, or rather Ares, the proper name for Gaia's sibling.

The idea of developing Mars as a colony has received surprisingly little attention except from science fiction writers. Our book was written as fiction although, as wisely observed by Brian Aldiss in his review, it was more a pamphlet, a serious idea in a fictional setting. We chose this format because of a chastening experience following the publication of a previous book, one about the great extinction of 65 million years ago when the great lizards and much of the rest of the biota perished. It was written as a popular science book, stimulated by the imaginative science of the Alvarez family and their collaborators, who attributed the extinction event to the impact of a large planetesimal. They supplied what seemed to us to be convincing evidence of such a collision, the discovery that iridium and other rare extraterrestrial elements were significantly more abundant at the boundary of the Cretaceous and Tertiary rocks. This is the place in the geological record that marks a large change in the populations and species of the living things that made the rocks. Shortly after its publication, the book was savagely criticized by paleontologists who wrote in those journals that set the scientific trend. Maybe their criticisms were necessary and the punishment was just. We should have, as travelers to an unfamiliar scientific territory, taken steps to learn its language and history and to have had the right visas and letters of introduction to the princes there—above all, to have been prepared for trouble in a land that was the home of greatest macho, *Tyrannosaurus*.

But we learnt our lesson, and wrote *The Greening of Mars* as fiction in the expectation that it would not be adversely criticized on points of intricate factual detail. Our book was intended as the scene for a series of imaginary, *gedanklich,* experiments on another planet. What if Mars, now a hopelessly barren desert, could be made fit for life? How could we then seed it and how would it develop? Neither of us expected it to be taken as more than entertainment. We should have known that everyone, or almost everyone, takes fiction much more seriously than fact. Just think for a moment: if you want to know the sociology of Victorian England, you could read Marx, who was the first social scientist, but more likely, even if you are a Marxist, you will read Dickens. Within months of publication in 1984, our second book stirred far more serious attention than its light-hearted writing seemed to merit. Three scientific meetings on the topic of making a second home on Mars were held, and at one of them, Robert Haynes, a distinguished geneticist from Toronto, coined the word *ecopoiesis*—literally, "the making of a home"—for the practice of transforming an otherwise uninhabitable environment into a place fit for life to evolve naturally. I prefer it to the word *terraforming,* often used when considering this act for planets. Ecopoiesis is more general. Terraforming has the homocentric flavor of a planetary-scale technological fix.

A key step in the development of a new geophysiological system is the acquisition of some novel and inheritable activity by a single organism. It follows that the first act in the ecopoiesis of Mars would have to be made by an entrepreneur. It would be an opportunistic act for private selfish gain; the larger communal act of colonization would come later. Columbus, I think, was not the chairman of a committee, but I suspect that those who traveled later aboard the *Mayflower* were the members of one.

To make Mars a fit home for life we shall first have to make the planet comfortable for bacterial life. In the book, we proposed that this impossible and outrageous act, the changing of the environment of a whole planet, could only be done by a slightly disreputable entrepreneur; the type of man about whom it is said, "He never breaks the law but whenever he does something, legislation is needed to stop him from doing it again." People like this are needed to probe the boundaries and to do those things that are forbidden, things that are apparently too costly or are beyond the possibility of achievement by the well-meant but sometimes undesirable caution of the planned enterprise of governmental agencies.

The scenario of *The Greening of Mars* included therefore a buccaneering character called Argo Brassbottom; later in life, success induced a snobbish gentility that caused him to change his surname to Foxe. He was a dealer in surplus weapons, and had the notion that there must be money to be made from the disposal of the vast accumulation of large, out-of-date ICBMs and other military rocket vehicles. The nuclear warheads could be, and would be, reprocessed as plutonium plowshares or future swords under strict governmental control. But what of the rocket carcasses full of solid propellant? These could not safely be disassembled and reused but they could, without modification, be the key components of a private space program. Brassbottom, through his many contacts in the civil and military services of the West and East, soon found that there would indeed be a reward for disposing of these unwanted rockets. Then he had another bright idea. His main line of business was as an industrial scavenger, a human dung beetle who profited from the disposal of toxic wastes and other noxious products that we prefer not to notice. Why not, he thought, use the rockets to propel the toxic wastes right outside the Earth? Deep space could be a safe dumping place.

Moving as he did among the black markets of the world, he was well acquainted with those unscrupulous scientists who will supply their skills, for a fee, to political fanatics or criminals. One of these commented that the recent anxiety over the state of the ozone layer had led to legislation banning chlorofluorocarbon aerosol propellants. Maybe there was a surplus of these products that required their expensive enclosure in vast pressurized tanks. These gases are among the most harmless and benign of chemicals that enter the home. They are not flammable, nor are they toxic or noxious. They were banned because their presence in the atmosphere could deplete stratospheric ozone. Why not, thought Brassbottom, propel them out into deep space and be paid for so doing? It was not long before another scientist suggested sending them to Mars. The chlorofluorocarbons are 10,000 times more potent than carbon dioxide as greenhouse gases to absorb the infrared radiation that escapes from the Earth. On Mars this property might lift off the frozen atmosphere. Brassbottom was enough of a businessman to get title to develop Mars, realizing that the stocks of his Mars development company would boom should the planet get a temperate climate and so become potentially habitable. As a final step, with the help of

friends in the United Nations agencies he convinced the new
government of the small archipelago of New Ulster in the Indian
Ocean to participate in building a launch site for his rockets
on the temporarily quiescent volcanic island of Crossmaglen.
It was heralded as the space program of the underdeveloped
world. Earnest scientists who persisted in taking our fictional
scenario as if it were up for peer review have pointed out to
me that this would not have worked because the CFCs are
rapidly destroyed by solar ultraviolet, and that carbon tetrafluoride, which is not destroyed, should have been proposed instead.
Maybe they are right.

When you are building imaginary worlds in the spaces of
the mind, tiresome details such as the solidity of the planetary
foundations and the presence or absence of rising damp or dry
rot tend to be ignored. What counts is the position of the property
and the grand view across the untouched landscape. Neither
Mike Allaby nor I realized the extent to which our dream worlds
would be seen as real estates. It is essential therefore, before
any of us are carried away, to go back and re-examine our
book as if it were a prospectus and not a work of fiction. If
we are to avoid, even in the imagination, accusations of fraudu-
lent deception, we need to include also a report on the state
of Mars from an independent surveyor. By rights this should
have been the task of the two Vikings, but sadly their directors
were obsessed by another fictional dream, that of finding life
on Mars. They should have made the necessary, albeit dull,
measurements of the abundance of light elements in the surface
rocks, the ratio of hydrogen and deuterium in the atmosphere,
and the structure of the Martian crust; instead these were given
less attention than the feverish but pointless search for life.

So what do we know of Mars? The best and most readable
summary of the information gathered by the spacecraft that
orbited or landed on the Martian surface is Michael Carr's
splendid and beautifully illustrated book, *The Surface of Mars*.
It includes many photographs taken from orbiting spacecraft.
Mars is seen to resemble the Moon much more than the Earth.
Impact craters pockmark the surface and reveal a preserved

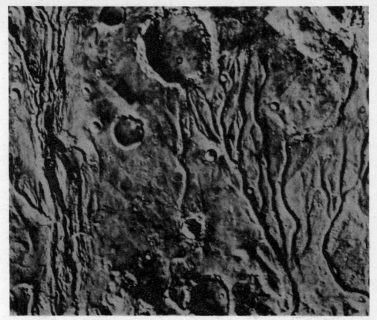

8.1 Water channels on Martian surface. The photographs from space show evidence of channels along which water may once have flowed early in the history of Mars.

chronicle of events going back to the planet's beginnings. This is in stark contrast to the Earth, where the ceaseless motions of the crust and the weathering by wind and water forever keep her face fresh and clean. Mars differs from the Moon in having an atmosphere, thin though it may be. It also has volcanoes, similar in form to those of the Hawaiian Islands but much larger. There are canyons and channels and dried-out river systems, suggesting that once long ago Mars had flowing water (see figure 8.1); there are polar caps that change their extent with the seasons; and there are clouds and dust storms in the thin remnants of its atmosphere.

Mars may seem to be dry, but much water has outgassed from the interior during the planet's history. The total quantity is thought to be somewhere between 12 and 25 million cubic kilometers (2.6 to 5.2 million cubic miles), enough to provide an ocean between 80 and 160 meters deep over the whole planet were it a smooth round sphere, or about 200 meters deep for a distribution of land and sea as on the Earth.

Michael McElroy of Harvard University has drawn on data for the isotopic composition of the element oxygen in the Martian atmosphere to argue that there has been little loss of water to space despite the lesser gravitational pull of Mars. Surprisingly the same arguments, when applied to the element nitrogen, lead to the conclusion that Mars has lost a large proportion of its nitrogen to space. There is strong evidence of massive floods and enough water to have produced river valleys nearly 1,000 kilometers long, but this was in the remote past. Where has all this water gone? According to Michael Carr's summary of the available evidence, most of the water now present is likely to be permafrost extending as deep as 1 or 2 kilometers below the surface. Layers of brine, with a freezing point as low as $-20°C$, may underlay the ice. In addition, the polar regions may overlay domes of ice.

That, then, is the present consensus among scientists about Mars. There may seem to be plenty of water, but for various reasons it would be as inaccessible to a colonizing biota as the water below the desert of Australia. In addition, to melt and vaporize the water deep below the surface, heat must be transferred from above. Heat transfer through a surface layer of dust can be astonishingly slow; if limited to the process of simple diffusion it could take millions of years to melt the subsurface ice. This may be a pessimistic conclusion. Frazer Fanale and his colleagues at the Jet Propulsion Laboratory have proposed that the movement of carbon dioxide gas through the rock dust will exert a flushing action and so transfer water to the surface. Changes of atmospheric pressure due to the condensation and evaporation of carbon dioxide are the driving force for this motion. But, on a human time scale, the act of ecopoiesis to bring Mars to the point of seeding could still be unbearably slow.

Before we take the drastic step of selling up our home on Earth, we need a great deal more information about our future home than was given by the Viking survey report. We need to know what could be the worst in store for us, and indeed for Mars itself, as a place for ecopoiesis.

If you look again at the lunar-like surface of Mars you will see that the channels and flow systems, which so strongly suggest the presence of water, are ancient indeed; almost all date to the period before 3.5 eons ago when planetesimal impacts were more frequent. Mars may have had a thicker atmospheric greenhouse and a warmer climate; also, there may have been heating from the impacts. Four eons ago, the Sun was at least 25 percent less luminous than now. If Mars is frozen now, a thick blanket would have been needed then to sustain an atmosphere and flowing water. Since those distant times, the Sun has warmed and there have been more large planetesimal impacts, although less frequent than in the early days. In spite of this no signs of further water flows are seen. The present coventional wisdom that envisages an ocean of frozen water 100 meters thick may be wrong. Not enough account has been taken of the probability that Mars, like the Earth, was originally rich in chemical substances that react with water to form hydrogen that escapes to space. The water may once have been there, but the escape of hydrogen left oxygen behind, not as free oxygen, but chemically bound in nitrates, sulfates, and iron oxides.

Consider the state of Mars 3.5 eons ago. This would be just after the planetesimals had rained down so immoderately and turned to rock and dust the entire planetary surface to a depth of at least 2 kilometers, a process that the planetologists coyly call "gardening." At that time the Earth was reducing; the environment was rich in those chemical compounds of iron and sulfur that have a considerable capacity to react with oxygen. There is no reason to believe that Mars was different. In addition, those early rocks had a considerable capacity to react with carbon dioxide. A 2-kilometer layer of powdered rock derived from basic basalt has the capacity to react with about 600 meters of water and carbon dioxide (3 bars), enough to make the surface atmospheric pressure of Mars three times greater than that on Earth now. Could this account for the thin atmosphere and aridity of Mars now? The abundant water that flowed 3.5 eons ago could have reacted with the ferrous iron of the rock dust, releasing the hydrogen it carried so that it escaped to space. It might be thought that the gas-solid reactions of weathering would be too slow to have removed much oxygen and carbon dioxide. This would be true of the present conditions on Mars; but if free water were present, much of the ferrous iron and sulfides could have been dissolved by the water, or dispersed as a fine slurry, hastening both the reactions themselves and the process of rock digestion. The oxidized state of Mars now, which gives the planet its deep red color, may be only skin deep. But until another surveyor, like Viking, goes there and tests the rocks at depth we cannot be sure that there is an atmosphere and water waiting for us.

It is worth reminding ourselves how the Earth avoided the same fate and why we also are not now desiccated. The carbon dioxide originally in the atmosphere has nearly all gone to form limestones and carbonaceous sedimentary rocks. Vast quantities of sulfides and ferrous iron have been oxidized, and the oxygen retained by this process may well have originally been associated with hydrogen in water. The Earth was saved from drying out by the abundance of its water, and by the presence of Gaia, who acts to conserve water. Mars could soon have lost its meager first water, and that may be why those channels are so ancient and why there is so little evidence of bulk water of recent origin. Mars may be irredeemably arid, and what little water is left may be deep below the surface in aquifers as salt and bitter as the Dead Sea. For most living organisms, saturated brine is hardly better than no water.

I must confess a personal intuition that Mars is nearer to a state of aridity. I cannot so easily envisage Mars as some potentially lush but deep-frozen sleeping beauty of a planet that waits to have the breath of life blown in from Earth. But fairy stories are much more entertaining than a dry-as-dust view of Mars; so let us accept the current scientific consensus that predicts abundant water and carbon dioxide waiting to be thawed, and let us use this pleasing model as the inspiration for our ecopoietic colonists. There remains only the questions of how we move in and what we should do to prepare the garden for planting.

If you were to visit Mars on a sunny summer afternoon in latitudes corresponding to those of Buenos Aires or Melbourne you might be surprised by the warmth of the climate. Daytime temperatures could be as high as 70°F. If only the air were breathable, it would be a shirt-sleeve environment. But on other days it might be below freezing. And always when the Sun went down the temperature would fall, with frightening rapidity, to reach −120°F by midnight; cold enough for solid carbon dioxide to form a frost of dry ice at the bottom of the valleys or depressions.

The ground beneath your feet would seem like desert on the Earth. But this would be an illusion, for few deserts anywhere on Earth are devoid of life. There is almost everywhere on Earthly deserts a thin cover of bacterial growth called the *desert pavement*. There is no soil on Mars, only a lifeless mix of rocks of all sizes from dust to boulders that has been given, almost onomatopoeically, that dry, harsh name, regolith (shown in figure 8.2). Mars is not yet ready for life; it is not only inhospitable to any form of life, it is also poisonous and destructive to organic matter. The air at the surface of Mars is in a chemical state like that of the stratosphere above the Earth. If the stratospheric air 10 miles above our heads could be compressed without changing its composition, we could not breathe it. Ozone is present there at 5 parts per million. Ozone may shield us from solar ultraviolet radiation, but at this abundance it is painful and soon lethal to breathe. The surface of Mars after a planetary lifetime exposed to such an atmosphere is rich in exotic chemicals, such as pernitric acid, that can rapidly destroy seeds, bacteria, or indeed almost all organic matter. Mars is no place for gardening.

The highly oxidized surface on Mars today means that life cannot spontaneously develop there. Unlike the Archean Earth, the organic precursors of living matter would not have the chance to accumulate and assemble. The only route for ecopoiesis is, first, for us to change the environment until it is suitable for

8.2 Regolith seen from Viking Lander. Mars has no soil—soil is the structured active surface of a living planet, regolith is rubble spread on the surface of a dead planet.

life and, then, either to allow it to evolve spontaneously or to seed the planet. If we achieve the environmental maturation, I cannot believe that we would have the patience to leave Mars to develop life alone. Someone would seed it, if only by accident.

Planetary life needs an operating system like Gaia, otherwise it is vulnerable to any change in its environment that could happen as a result either of its own evolution or of an external disaster such as the all-too-frequent impact of planetesimals. I do not believe that sparse life, existing only in a few oases on a planet, is viable. Such a system is incomplete; unable to control its environment and powerless to resist adverse change. It follows that, even if we sprayed every bit of the planet's surface with every species of microorganism, we could not bring Ares to life. Some organisms might survive and even grow for a brief spell, but there would be no invasion, no infection with the rapid spread of life to take over and control the planet. I find it unusual that otherwise capable organizations like NASA should strive so hard to sterilize their spacecraft when they well know that Mars itself is a great sterilizer. They also know that were the same craft to land unsterilized on the much more hospitable terrain of the antarctic ice cap or the Australian desert, their small complement of microbial passengers would have no chance of establishing a permanent home there.

Parts of Mars may now have equable temperature on sunny afternoons, but this does not mean that little needs be done to bring it alive. When life began on Earth, the heat received from the Sun was 60 percent greater than that now warming Mars. There was abundant water on Earth and a dense enough atmosphere to provide a comfortable climate. The only thing in Mars' favor is that it is darker than the Earth and absorbs more of the sunlight falling on it. But this advantage is only for its present state; once water is set free it will evaporate to form clouds and snow cover. This will increase the albedo of Mars so that it reflects to space the heat that it might otherwise have gained. Mars by itself may never be able to provide the conditions needed to start and sustain life, not even in a billion years' time when the Sun is hotter and what is left of the Martian air and water has been set free.

What can we then do to start Mars on the evolutionary course that would eventually bring it to a condition like that on Earth now and so become our second home? First, the Martian environment must be changed sufficiently to allow spontaneous growth and spread of microorganisms over a large proportion of the planetary surface. At first glance the notion of planetary engineering, the ecopoiesis of a planet, seems a grandiose impertinence. But it is not so impertinent if Mars is a deep-frozen planet needing only to be thawed; moreover this is the consensus view among planetary scientists, who report that as much as 2 atmospheres pressure of carbon dioxide and enough water to cover the planet to a depth of 100 meters or so have outgassed from its interior over the past 4 eons. If we accept this conclusion, then we could think of Mars as poised on the edge of a cliff of environmental stability; a small push may be enough to change it to a state much more suited to life.

In his book on Mars, Michael Carr discusses the possibility

that liquid water exists in aquifers beneath the surface of the planet, also the likelihood that such water might be salt. It is often forgotten that the stable state of the element nitrogen is as the nitrate ion, dissolved in water. On Earth, nitrate is formed continuously by high-energy processes (fires, lightning, and nuclear radiation) in the atmosphere. It quickly reaches the ground in rain, and the biota equally promptly return it to the atmosphere as nitrogen gas. There is no life on Mars, and I have often wondered if most of the nitrogen is there as nitrate dissolved in the brines. Or maybe there are vast salt deposits, evaporite beds, left after the ancient water flows dried out. Nitrate and nitrite locked up in these deposits could also account for the relative lack of nitrogen in the present Martian atmosphere.

It will take another Viking to find the answers to these questions, and for now we can only speculate about what changes would have to occur to convert the present infertile Mars into a seedbed for planetary life. That is why Mike Allaby and I chose to write our tale of Martian ecopoiesis as fiction, and to warm Mars by projecting surplus CFCs from Earth. I have my doubts about whether enough of these powerful greenhouse gases could be sent, but this idea was intended to titillate the imagination of those who might want to convert Mars by some other means, rather than as a serious engineering proposal. I have often found in my practice as an inventor that a slightly wrong or incomplete invention is more attractive to engineers than one that is a *fait accompli*. In any event it seems greedy to attempt more than one's proper part of a project, to take from others the chance to exercise their special skills and artistry.

Instead of sending CFCs to Mars expensively by spacefreight delivery as proposed in our book, someone may design an automatic plant to manufacture them on Mars from indigenous materials. If the Martian brines exist, and can be tapped, it should be no great task to synthesize fluorocarbons and other potential greenhouse gases, such as carbon tetrafluoride, using the salts of the brines and atmospheric carbon dioxide as the raw materials. It would require a moderate-sized nuclear power plant. Maybe environmentalists would be glad to see one shipped to Mars instead of sited here on Earth. If nitrate and nitrite are present in the brines, then these will provide a convenient local source of both oxygen and nitrogen. Not enough to change the atmosphere, but plenty for early explorers and technicians to breathe in their enclosed habitats.

We have proposed the warming of Mars by sending greenhouse gases there; would it work? The basic mechanism of the greenhouse effect looks simple enough, but to calculate the temperature rise corresponding to a stated increase in carbon dioxide is far from simple. On a planetary scale many other things must be taken into account, including the reflection of sunlight by clouds and ice cover; the transport of heat by air movement and by the evaporation and condensation of water; and atmospheric and ocean structure. Not surprisingly, these calculations require the help of the largest computers that are available, and even they are inadequate. So far as I am aware, no models have included the dynamic responsive feedback from the biota. The Martian greenhouse effect is likely to be a great deal easier to calculate—or at least it will be in the first stages before enough water has evaporated to introduce cloudiness, snow cover, and water vapor. Cloud and ice both are white and sunlight-reflecting. Broadly speaking, ice has the opposite effect of carbon dioxide and causes cooling; clouds can either heat or cool according to their form and altitude. To complicate the problem further, water vapor absorbs infrared, and its presence amplifies the heating effect of carbon dioxide.

The idea of warming Mars by introducing CFCs into the atmosphere depends upon a set of favorable coincidences. First, there is a broken pane in the greenhouse. Neither carbon dioxide nor water vapor are effective absorbers of infrared at wavelengths between 8 and 14 micrometers, and a fair amount of heat radiates away to space from the planetary surface and atmosphere at these wavelengths. The CFCs absorb intensely in this region and serve as a new pane of glass, still transparent to sunlight but opaque in what previously was a gap in the infrared. Second, greenhouse gases have a way of amplifying one another's effects. It is not commonly known outside meteorology that the carbon dioxide greenhouse depends mainly on the infrared absorption by water vapor. Carbon dioxide does absorb infrared radiation, but not at the same wavelengths or as strongly as does water vapor. An increase of carbon dioxide will cause some warming and this in turn will increase the water vapor content of the air. The increased water vapor increases the warming and so amplifies the smaller effect of carbon dioxide. On Mars there will be a double amplification. The CFCs will warm the surface a little, this will lift off carbon dioxide and so increase the warming, which will in turn evaporate water and still further warm the planet. This is why it may be possible, using a practical quantity of these strange chemicals, to change the climate of a whole planet. We cannot say, until the modeling is done, how much CFCs would be needed. It might be as little as 10,000 tons or more than one million tons. If it as large as the latter, Brassbottom's enterprise would not succeed; it would, however, still be within the capacity of an automated chemical plant shipped to Mars with the purpose of synthesizing these or other greenhouse gases from indigenous materials.

The success or failure of ecopoiesis for Mars is likely to depend on how much carbon dioxide and how much water is there in an available form. With a dense carbon dioxide atmosphere, 2 bars or more, a tolerable climate is likely. With less carbon dioxide, a great deal will depend on the distribution of water and on the effect of snow and clouds on the planetary albedo. In other science fiction scenarios water has been transported to Mars as asteroids of ice, taken from their frigid orbits far from the Sun. Simple calculations show the impracticality of this notion without some incredible new motive power. An asteroid of pure ice, 200 miles in diameter, is needed to equal the quantity of water now thought to be on Mars. Few would be prepared to take on the contract for moving it there.

When the CFCs have done their job of lifting an atmosphere from the previously frozen surface, what world do we have? Let us assume, for a start, that we have a planet with an atmosphere of between 0.5 and 2 bars pressure and composed almost wholly of carbon dioxide. The climate is still cold by Earthly standards but the diurnal fluctuations are less extreme; at low altitudes in the tropical regions the night frosts are no longer as frequent or severe. Most important, enough water has evaporated for precipitation to occur in some regions. The surface is still regolith but no longer highly oxidizing; the lethal pernitric acid and other stratospheric oxidants have moved up in the atmosphere to those high-altitude regions where they exist on Earth.

I can only guess at the ecosystem that could survive in such an environment. It would be unlikely to include the land plants and animals, at least not initially. The first life on Earth was the prokaryotic microorganisms, and their descendants still flourish in the soil. Our first objective would be to introduce a microbial ecosystem that could convert the regolith into topsoil, and at the same time to introduce surface-dwelling photosynthetic bacteria. These could provide the food, energy, and raw materials for the bulk of the ecosystem dwelling below the surface. If we could arrange that the photosynthesizers be colored dark, they would absorb the Sun's warmth and so be warmer than their surroundings. On a local scale this is like the advantage possessed by dark daisies on Daisyworld; it could encourage the ecosystem that they were a part of to spread across the Martian surface. If this happened the climate might tend towards homeostasis, at first by regions and finally globally.

There are other ways available to the biota for regulating climate in addition to the control of albedo. Probably most important is the regulation of the composition of the atmospheric gases. The first act of ecopoiesis was to build an artificial greenhouse made of CFC gases at a few parts per billion in the air. In the early life of Ares, the control of the CFC emissions would still be available from the human colonists. This may be especially important if the atmospheric carbon dioxide is significantly reduced or if snow and cloud cover increase the planetary albedo. There are two ways that carbon dioxide might be removed in significant amounts. The first is if the life is so successful in its spread that it splits large quantities of the gas into carbonaceous organic matter and free oxygen. The second is by the reaction of carbon dioxide with calcium silicate rocks to form carbonates and silicic acid. The first reactions would release free oxygen, which might accumulate in the air; it might be that the rocks of the regolith and the water of the Martian brines contain a fair quantity of materials that scavenge oxygen, such as the element iron in its ferrous form. In any case, the first oxygen to appear in the atmosphere will be too dilute to permit the easy reoxidation of the surplus organic matter produced by photosynthesis. The surplus carbon of the dead photosynthesizers would be reoxidized by other organisms of the bacterial ecosystem using as oxidants the sulfate and nitrate of the regolith. This would return carbon dioxide, nitrogen, and nitrous oxide to the air. Before long, however, the soil of Mars would be tending towards a state where there would be insufficient oxidants as nonrenewable resources to sustain the reoxidation of carbonaceous matter and the return of carbon dioxide to the air. When this point was reached on the early Earth in the Archean, it opened for exploitation a giant niche of surplus organic matter. It was then, I think, that the methanogens evolved to take an opportunistic advantage of this gift from the photosynthesizers. In doing so they converted the organic matter to a mixture of methane and carbon dioxide. Methane is also a greenhouse gas, so the potentially disastrous cooling that might otherwise have occurred was avoided.

Already in this brief discussion we have postulated the need for photosynthesizers, nitrate and sulfate reducers, and methanogens. All are normal inhabitants in a sample of soil from almost anywhere on Earth. Aerobic and anaerobic ecosystems peacefully coexist with their respective territories segregated on a vertical basis so that the oxygen-tolerant are at the surface and the anaerobes at the lowest point of the soil. The soil is a complex and intricate assembly, and diverse in its population of species. Successfully establishing the bacterial ecosystem of soil in the Martian regolith is not a matter of finding, or making by genetic engineering, species that will grow there; it is a matter of changing Mars to a state where the microbial ecosystems of the Earth can flourish and convert the regolith to soil. But that is still only the start, for if Mars is to become a self-sustaining system it is necessary for the organisms and their environment to become as tightly a coupled system as they are on Earth. The acquisition of planetary control can come only from the growing together of life and its environment until they are a single and indivisible system.

One family in a dwelling does not make a village, still less does it constitute a city with a self-sustaining infrastructure. In the same way there is a critical mass of biota needed for planetary homeostasis, the size of which depends mainly upon how much effort is needed to sustain homeostasis and how large are the perturbations likely to take place. Simple models, derived from Daisyworld, suggest that a stable system requires at least 20 percent cover of the planetary surface if the commoner perturbations are to be withstood. These would be changes in the intensity of sunlight, planetesimal impacts, internal disturbances from the evolution of species that adversely affected the environment, or the exhaustion of some essential resource. If Ares is to grow strong he will need to cover more than just a few oases of the Martian desert.

In romantic novels, the excitement is placed before the wedding. That is great for entertainment, but it is no guide for successful married life. So it is with ecopoiesis; the physical and chemical conversion of Mars would be an incredible feat of engineering, a great and enduring saga. In contrast, the nursing of the infant planetary life, though fulfilling, would seem an anticlimax. Great patience and love would need be given to the unremitting task of nurture and the daily guidance of the newborn planetary life until it could, by itself, sustain homeostasis.

Thoughts of Gaia will always be linked with space exploration and Mars, for in a sense Mars was her birthplace. Rusty Schweickart and his fellow astronauts have shared with us their revelation on looking back at the Earth from the distance of the Moon; their realization that it was their home. In a lesser, but still significant way, our vicarious view of the planets of the Solar System seen through the splendid eyes of the Voyager and other spacecraft has touched our minds and set in motion the locked plates of the Earth sciences.

Lord Young, prominent for his work towards the founding of the open university in the United Kingdom, has been so moved by the idea of bringing life to Mars that he has formed the Argo Venturers to think and act towards this end. He believes that the prospect of colonizing Mars, even before or without its final achievement, is a powerful source of inspiration. I share his view, and think that the contemplation of the daunting difficulties of bringing Ares to life may help us better to understand the awful consequences of so damaging Gaia that we have to take on the never-ceasing responsibility of keeping the Earth a fit place for life, a service now provided for free.

9

God and Gaia

Gaia, mother of all, I sing, oldest of gods,
Firm of foundation, who feeds all creatures living on Earth,
As many as move on the radiant land and swim in the sea
And fly through the air—all these does she feed with her bounty.
Mistress, from you come our fine children and bountiful harvests,
Yours is the power to give mortals life and to take it away.
  J. DONALD HUGHES, *Gaia: An Ancient View of Our Planet*

Photographs, like biographies, often reveal more of the artist than of the subject. Maybe this is why passport photographs, taken in mechanically operated booths, look so lifeless. How could a mere machine capture the soul of its subject, stiffly sitting and gazing into the blind eye of the camera? Trying to write about God and Gaia, I share some of the limitations of a mechanical camera, and I know that this chapter will show more about myself than about my subjects. So why try?

When I wrote the first book on Gaia I had no inkling that it would be taken as a religious book. Although I thought the subject was mainly science, there was no doubt that many of its readers found otherwise. Two-thirds of the letters received, and still coming in, are about the meaning of Gaia in the context of religious faith. This interest has not been limited to the laity; a most interesting letter came from Hugh Montefiore, then Bishop of Birmingham. He asked which I thought came first,

life or Gaia. My attempts to answer this question led to a corre-
spondence, reported in a chapter of his book *The Probability
of God*. I suspect that some cosmologists are similarly visited
by enquiries from those who imagine them to be at least on
nodding terms with God. I was naive to think that a book
about Gaia would be taken as science only.

So where do I stand about religion? While still a student I
was asked seriously, by a member of the Society of Friends, if
I had ever had a religious experience. Not understanding what
he meant, imagining that he referred to a manifestation or a
miracle, I answered no. Looking back from 45 years on, I now
tend to think that I should have said yes. Living itself is a
religious experience. At the time, however, the question was
almost meaningless because it implied a separation of life into
sacred and secular parts. I now think that there can be no
such division. In any relationship there are high points of delight,
as well as pitfalls in the great plain of contentment. For me
one high point came when I was asked by Jim Morton, the
Dean of the Cathedral Church of St. John the Divine in New
York, to serve as a participant in a religious celebration. I still
recall with wonder being part of that colorful procession, with
him and other clerics, dressed in medieval costume. The music
of the choir singing "Morning Is Broken" seemed to take on a
new significance in the ambience of that sacred place. It was
a sensual experience, but to me that does not make it less religious.

My thoughts about religion when a child grew from those
of my father and the country folk I knew. It was an odd mixture,
composed of witches, May trees, and the views expressed
by Quakers, in and outside the Sunday school at a Friends'
meeting house. Christmas was more of a solstice feast than
a Christian one. We were, as a family, well into the present
century, yet still amazingly superstitious. So ingrained was my
childhood conditioning about the power of the occult that
in later life it took a positive act of will to stop touching
wood or crossing fingers whenever some hazard was to
be faced. Christianity was there not so much as a faith,

rather as a set of sensible directions on how to be good.

When I first saw Gaia in my mind I felt as an astronaut must have done as he stood on the Moon, gazing back at our home, the Earth. The feeling strengthens as theory and evidence come in to confirm the thought that the Earth may be a living organism. Thinking of the Earth as alive makes it seem, on happy days, in the right places, as if the whole planet were celebrating a sacred ceremony. Being on the Earth brings that same special feeling of comfort that attaches to the celebration of any religion when it is seemly and when one is fit to receive. It need not suspend the critical faculty, nor can it prevent one from singing the wrong hymn or the right one out of tune.

That is only what I feel about Gaia. What about God? I am too committed to the scientific way of thinking to feel comfortable when enunciating the Creed or the Lord's Prayer in a Christian Church. The insistence of the definition "I believe in God the Father Almighty, Maker of Heaven and Earth" seems to anesthetize the sense of wonder, as if one were committed to a single line of thought by a cosmic legal contract. It seems wrong also to take it merely as a metaphor. But I respect the intuition of those who do believe, and I am moved by the ceremony, the music, and most of all by the glory of the words of the prayer book that to me are the nearest to perfect expression of our language.

I have kept my doubts in a separate place for too long. Now that I write this chapter, I have to try somehow to explain, to myself as well as to you, what is my religious belief. I am happy with the thought that the Universe has properties that make the emergence of life and Gaia inevitable. But I react to the assertion that it was created with this purpose. It might have been; but how the Universe and life began are ineffable questions. When a scientist colleague uses evidence about the Earth eons ago to explain his theory of the origins of life it stirs a similar sense of doubt. How can the events so long ago that led to the emergence of anything so intricate as life be treated as a fact of science? It is human to be curious about antecedents, but expeditions into the remote past in search of origins is as supremely unimportant as was the hunting of the snark. The greater part of the information about our origins is with us here and now; so let us rejoice in it and be glad to be alive.

At a meeting in London recently, a wise man, Dr. Donald Braben, asked me: "Why do you stop with the Earth? Why not consider if the Solar System, the Galaxy, or even the Universe is alive?" My instant answer was that the concept of a living Earth, Gaia, is manageable. We know that there is no other life in this Solar System, and the nearest star is utterly remote. There must be other Gaias circling other docile long-lived stars but, curious though I may be about them and about the Universe, these are intangible—concepts for the intellect, not the senses. Until, if ever, we are visited from other parts of the Universe we are obliged to remain detached.

Many, I suspect, have trodden this same path through the mind. Those millions of Christians who make a special place in their hearts for the Virgin Mary possibly respond as I do. The concept of Jahweh as remote, all-powerful, all-seeing is either frightening or unapproachable. Even the sense of presence of a more contemporary God, a still, small voice within, may not be enough for those who need to communicate with someone outside. Mary is close and can be talked to. She is believable and manageable. It could be that the importance of the Virgin Mary in faith is something of this kind, but there may be more to it. What if Mary is another name for Gaia? Then her capacity for virgin birth is no miracle or parthenogenetic aberration, it is a role of Gaia since life began. Immortals do not need to reproduce an image of themselves; it is enough to renew continuously the life that constitutes them. Any living organism a quarter as old as the Universe itself and still full of vigor is as near immortal as we ever need to know. She is of this Universe and, conceivably, a part of God. On Earth she is the source of life everlasting and is alive now; she gave birth to humankind and we are a part of her.

This is why, for me, Gaia is a religious as well as a scientific concept, and in both spheres it is manageable. Theology is also a science, but if it is to operate by the same rules as the rest of science, there is no place for creeds or dogma. By this I mean theology should not state that God exists and then proceed to investigate his nature and his interactions with the Universe and living organisms. Such an approach is prescriptive, presupposes his existence, and closes the mind to such questions as: What would the Universe be like without God? How can we use the concept of God as a way to look at the Universe and ourselves? How can we use the concept of Gaia as a way to understanding God? Belief in God is an act of faith and will remain so. In the same way, it is otiose to try to prove that Gaia is alive. Instead, Gaia should be a way to view the Earth, ourselves, and our relationships with living things.

The life of a scientist who is a natural philosopher can be deeply religious. Curiosity is an intimate part of the process of loving. Being curious and getting to know the natural world leads to a loving relationship with it. It can be so deep that it cannot be articulated, but it is nonetheless good science. Creative scientists, when asked how they came upon some great discovery, frequently state, "I knew it intuitively, but it took several years work to prove it to my colleagues." Compare that statement with this one by William James, the nineteenth-century philosopher and psychologist, in *The Varieties of Religious Experience:*

The truth is that in the metaphysical and religious sphere, articulate reasons are cogent for us only when our inarticulate feelings of reality have already been impressed in favor of the same conclusion. Then, indeed, our intuitions and our reason work together, and great world ruling systems, like that of the Buddhist or of the Catholic philosophy, may grow up. Our impulsive belief is here always what sets up the original body of truth, and our articulately verbalised philosophy is but a showy translation into formulas. The unreasoned and immediate assurance is the deep thing in us, the reasoned argument is but a surface exhibition. Instinct leads, intelligence does but follow.

This was the way of the natural philosophers in James Hutton's time in the eighteenth century and is still the way of many scientists today. Science can embrace the notion of the Earth as a superorganism and can still wonder about the meaning of the Universe.

How did we reach our present secular humanist world? In times that are ancient by human measure, as far back as the earliest artifacts can be found, it seems that the Earth was worshipped as a goddess and believed to be alive. The myth of the great Mother is part of most early religions. The Mother is a compassionate, feminine figure; spring of all life, of fecundity, of gentleness. She is also the stern and unforgiving bringer of death. As Aldous Huxley reminds in *The Human Experience:*

> In Hinduism, Kali is at once the infinitely kind and loving mother and the terrifying Goddess of destruction, who has a necklace of skulls and drinks the blood of human beings from a skull. This picture is profoundly realistic; if you give life, you must necessarily give death, because life always ends in death and must be renewed through death.

At some time not more than a few thousand years ago the concept of a remote master God, an overseer of Gaia, took root. At first it may have been the Sun, but later it took on the form we have with us now of an utterly remote yet personally immanent ruler of the Universe. Charlene Spretnak, in her moving and readable book, *The Spiritual Dimensions of Green Politics,* attributes the first denial of Gaia, the Earth goddess, to the conquest of an earlier Earth-centered civilization by the Sun-worshipping warriors of the invading Indo-European tribes.

Picture yourself as a witness of that decisive moment in history, that is, as a resident of the peaceful, artful, Goddess-oriented culture in Old Europe. (Don't think "matriarchy"! It may have been, but no one knows, and that is not the point.) It is 4,500 BC. You are walking along a high ridge, looking out across the plains to the east. In the distance you see a massive wave of horsemen galloping towards your world on strange, powerful animals. (The European ancestor of the horse had become extinct.) They brought few women, a chieftain system, and only a primitive stamping technique to impress their two symbols, the sun and a pine tree. They moved in waves first into southeastern Europe, later down into Greece, across all of Europe, also into the Middle and Near East, North Africa and India. They brought a sky god, a warrior cult, and patriarchal social order. *And that is where we live today*—in an Indo-European culture, albeit one that is very technologically advanced.

The evolution of these horsemen to the modern men who ride their infinitely more powerful machines of destruction over the habitats of our partners in Gaia seems only a small step. The rest of us, in the cozy, comfortable hell of urban life, care little what they do so long as they continue to supply us with food, energy, and raw materials and we can continue to play the game of human interaction.

In ancient times, belief in a living Earth and in a living cosmos was the same thing. Heaven and Earth were close and part of the same body. As time passed and awareness grew of the vast distances of space and time through such inventions as the telescope, the Universe was comprehended and the place of God receded until now it hides behind the Big Bang, claimed to have started it all. At the same time, as population increased so did the proportion forced to lead urban lives out of touch with Nature. In the past two centuries we have nearly all become city dwellers, and seem to have lost interest in the meaning of both God and Gaia. As the theologian Keith Ward wrote in the *Times* in December 1984:

It is not that people know what God is, and have decided to reject him. It seems that very few people even know what the orthodox traditional idea of God, shared by Judaism, Islam and Christianity, is. They have not the slightest idea what is meant by the word God.

It just has no sense or possible place in their lives. Instead they either invent some vague idea of a cosmic force with no practical implications at all; or they appeal to some half-forgotten picture of a bearded super-person constantly interfering with the mechanistic laws of Nature.

I wonder if this is the result of sensory deprivation. How can we revere the living world if we can no longer hear the bird song through the noise of traffic, or smell the sweetness of fresh air? How can we wonder about God and the Universe if we never see the stars because of the city lights? If you think this to be exaggeration, think back to when you last lay in a meadow in the sunshine and smelt the fragrant thyme and heard and saw the larks soaring and singing. Think back to the last night you looked up into the deep blue black of a sky clear enough to see the Milky Way, the congregation of stars, our Galaxy.

The attraction of the city is seductive. Socrates said that nothing of interest happened outside its walls and, much later, Dr. Johnson expressed his view of country living as "One green field is like another." Most of us are trapped in this world of the city, an everlasting soap opera, and all too often as spectators, not players. It is something to have sensitive commentators like Sir David Attenborough bring the natural world with its visions of forests and wilderness to the television screens of our suburban rooms. But the television screen is only a window and only rarely clear enough to see the world outside; it can never bring us back into the real world of Gaia. City life reinforces and strengthens the heresy of humanism, that narcissistic devotion to human interests alone. The Irish missionary Sean McDonagh wrote in his book, *To Care for the Earth:* "The 20 billion years of God's creative love is either seen simply as the stage on which the drama of human salvation is worked out, or as something radically sinful in itself and needing transformation."

The heartlands of the great religions are now in the last bastions of rural existence, in the Third World of the tropics. Elsewhere God and Gaia that once were joined and respected are now divorced and of no account. We have, as a species, almost resigned from membership in Gaia and given to our cities and our nations the rights and responsibilities of environmental regulation. We struggle to enjoy the human interactions of city life yet still yearn to possess the natural world as well. We want to be free to drive into the country or the wilderness without polluting it in so doing; to have our cake and eat it. Human and understandable such striving may be, but it is illogical. Our humanist concerns about the poor of the inner cities or the Third World, and our near-obscene obsession with death, suffering, and pain as if these were evil in themselves—these thoughts divert the mind from our gross and excessive domination of the natural world. Poverty and suffering are not sent; they are the consequences of what we do. Pain and death are normal and natural; we could not long survive without them. Science, it is true, assisted at the birth of technology. But when we drive our cars and listen to the radio bringing news of acid rain, we need to remind ourselves that we, personally, are the polluters. We, not some white-coated devil figure, buy the cars, drive them, and foul the air. We are therefore accountable, personally, for the destruction of the trees by photochemical smog and acid rain. We are responsible for the silent spring that Rachel Carson predicted.

There are many ways to keep in touch with Gaia. Individual humans are densely populated cellular and endosymbiont collectives, but clearly also identities. Individuals interact with Gaia in the cycling of the elements and in the control of the climate, just like a cell does in the body. You also interact individually in a spiritual manner through a sense of wonder about the natural world and from feeling a part of it. In some ways this interaction is not unlike the tight coupling between the state of the mind and the body. Another connection is through the powerful infrastructures of human communication and mass transfer. We as a species now move a greater mass of some materials around the Earth than did all the biota of Gaia before we appeared. Our chattering is so loud that it can be heard to the depths of the Universe. Always, as with other and earlier species within Gaia, the entire development arises from the activity of a few individuals. The urban nests, the agricultural ecosystems, good and bad, are all the consequences of rapid positive feedback starting from the action of an inspired individual.

A frequent misunderstanding of my vision of Gaia is that I champion complacance, that I claim feedback will always protect the environment from any serious harm that humans might do. It is sometimes more crudely put as "Lovelock's Gaia gives industry the green light to pollute at will." The truth is almost diametrically opposite. Gaia, as I see her, is no doting mother tolerant of misdemeanors, nor is she some fragile and delicate damsel in danger from brutal mankind. She is stern and tough, always keeping the world warm and comfortable for those who obey the rules, but ruthless in her destruction of those who transgress. Her unconscious goal is a planet fit for life. If humans stand in the way of this, we shall be eliminated with as little pity as would be shown by the micro-brain of an intercontinental ballistic nuclear missile in full flight to its target.

What I have written so far has been a testament built around the idea of Gaia. I have tried to show that God and Gaia, theology and science, even physics and biology are not separate but a single way of thought. Although a scientist, I write as an individual, and my views are likely to be less common than I like to think. So now let me tell you something of what the scientific community has to say on this subject.

In science, the more discovered, the more new paths open for exploration. It is usual in science, when things are vague and unclear, for the path to be like that of a drunkard wandering in a zigzag. As we stagger back from what lastly dawns upon our befuddled wits is the wrong way, we cross over the true path and move nearly as far to the, equally wrong, opposite side. If all goes well, our deviations lessen and the path converges towards, but never completely follows, the true one. It gives a new insight to the old tag *in vino veritas*. So natural is this way to find the truth that we usually program our computers to solve problems too tedious to do ourselves by setting them to follow the same trial-and-error, staggering, stumbling walk.

The process is dignified and mystified by calling it "iteration," but the method is the same. The only difference is that, so quickly is it done, the eye never sees the fumbling.

We have lost the instinctive understanding of what life is and of our place within Gaia. Our attempts to define life are much in the stage of the drunkard's walk. The two opposing verges representing the extremes of iteration are illustrated by a splendid philosophical debate that has gone on for the past twenty years between the molecular biologists on the one side and the new school of thermodynamics on the other. Jacques Monod's *Chance and Necessity,* although first published in 1970, most clearly and beautifully conveys the clear, strong, and rigorous approach of solid science based firmly in a belief in a materialistic and deterministic Universe. The other verge is represented by those, like Erich Jantsch, who believe in a self-organizing Universe. It is concerned with the thermodynamics of the unsteady state of which dissipative structures such as flames, whirlpools, and life itself are examples. Although the participants are all well known and respected in the English-speaking world, most of this entertaining debate has gone on in French, so many of us have missed the fun.

The essence of this contest is a rerun of the ancient battle between the holists and the reductionists. As Monod reminds us:

Certain schools of thought (all more or less consciously or confusedly influenced by Hegel) challenge the value of the analytical approach to systems as complex as living beings. According to these holist schools which, phoenix like, are reborn in every generation, the analytic attitude (reductionist) is doomed to fail in its attempts to reduce the properties of a very complex organization to the "sum" of the properties of its parts. It is a very stupid and misguided quarrel which merely testifies to the holists' total lack of understanding of scientific method and the crucial role analysis plays in it. How far could a Martian engineer get if trying to understand an earthly computer, he refused on principle to dissect the machine's basic electronic components which execute the operation of propositional algebra.

These strong words were in the 1970 edition of *Chance and Necessity*. Maybe they are by now less extremely held, but they serve well to express what was and still is an important scientific constituency.

No one now doubts that it was plain, honest reductionist science that allowed us to unlock so many of the secrets of the Universe, not least those of the living macromolecules that carry the genetic information of our cells. But clear, strong, and powerful though it may be, it is not enough by itself to explain the facts of life. Consider Jacques Monod's Martian engineer. Would it have been sensible to have dashed in with a kit of tools and disassembled analytically the computer he found? Or would it have been better, as a first step, to have switched it on and questioned it as a whole system? If you have any doubts about the answer to this question then consider the thought that the hypothetical Martian engineer was an intelligent computer and the object he examined, you.

By contrast, in 1972 Ilya Prigogine wrote:

It is not instability but a succession of instabilities which allow the crossing of the no man's land between life and no-life. We start to disentangle only certain stages. This concept of biological order leads automatically to a more blurred appreciation of the role of chance and necessity to recall the title of the well-known work by Jacques Monod. Fluctuation which allows the system to depart from states near thermodynamic equilibrium represents the stochastic aspect, the part played by chance. Contrariwise, the environmental instability, the fact that the fluctuations will increase, represents necessity. Chance and necessity cooperate instead of opposing one another.

I wholly agree with Monod that the cornerstone of the scientific method is the postulate that Nature is objective. True knowledge can never be gained by attributing "purpose" to phenomena. But, equally strongly, I deny the notion that systems are never more than the sum of their parts. The value of Gaia in this debate is that it is the largest of living systems. It can be analyzed both as a whole system and, in the reductionist manner, as a collection of parts. This analysis need disturb neither the privacy nor the function of Gaia any more than would the movement of a single commensal bacterium on the surface of your nose.

Prigogine was not the first to recognize the inadequacies of equilibrium thermodynamics. He had many illustrious predecessors, among them the physical chemists J. W. Gibbs, L. Onsager, and K. G. Denbigh, who explored the thermodynamics of the steady state. But it was that truly great physicist, Ludwig Boltzmann, who pointed the way towards the understanding of life in thermodynamic terms. And it was by reading Schrödinger's book *What Is Life?* in the early 1960s that I first realized that planetary life was revealed by the contrast between the near-equilibrium state of the atmosphere of a dead planet and the exuberant disequilibrium of the Earth.

When we cross from the sharp clarity of the real world into that nightmare land of dissipating structures, what do we learn that makes the next staggering lurch less erroneous than the last? I have gained from Prigogine's world view a confirmation of a suspicion that time is a variable much too often ignored. In particular, many of the apparent contradictions between these two schools of thought seem to resolve if viewed along the time dimension instead of in space. We have evolved from the world of simple molecules through dissipative structures to the more permanent entities that are living organisms. The further we go from the present, either into the past or the future, the greater the uncertainty. Darwin was right to dismiss thoughts about the origins of life; as Jerome Rothstein has said, the restrictions of the second law of thermodynamics prevent us from ever knowing about the beginning or the end of the Universe.

In our guts and in those of other animals, the ancient world of the Archean lives on. In Gaia, also, the ancient chaotic world of dissipating structures that preceded life still lives on. A recent and relatively unknown discovery of science is that the fluctuations at every scale from viscosity to weather can be chaotic. There is no complete determinism in the Universe; many things are as unpredictable as a perfect roulette wheel. An ecologist colleague of mine, C. S. Holling, has observed that the stability of large-scale ecosystems depends upon the existence of internal chaotic instabilities. These pockets of chaos in the larger, stable Gaian system serve to probe the boundaries set by the physical constraints to life. By this means the opportunism of life is insured, and no new niche remains undiscovered. For example, I live in a rural region surrounded by farmers who keep sheep. It is impressive how adventurous young lambs, through their continuous probing of my boundary hedges, can find their way through onto the richer, ungrazed land on my side. The behavior of young men is not so different.

My reason for wandering onto the battlefield of the war between holists and reductionists was to illustrate how polarized is science itself. Let me conclude this digressionary visit and return to the theme of this chapter, God and Gaia. And let me start by reminding you of Daisyworld—a model which is reductionist and holistic at the same time. It was made to answer a criticism of Gaia, that it was teleology. The need for reduction arose because the relationships between all the living things on Earth in their countless trillions and the rocks, the air, and the oceans could never be described in full detail by a set of mathematical equations. A drastic simplification was needed. But the model with its closed loop cybernetic structure was also holistic. This also applies to ourselves. It would be pointless to attempt to disentangle all the relationships between the atoms within the cells that go to make up our bodies. But this does not prevent us from being real and identifiable, and having a life span of at least 70 years.

We are also in an adversary contest between our allegiance to Gaia and to humanism. In this battle, politically minded humanists have made the word "reductionist" pejorative, to discredit science and to bring contumely to the scientific method. But all scientists are reductionists to some extent; there is no way to do science without reduction at some stage. Even the analyzers of holistic systems, confronted with an unknown system, do tests, such as perturbing the system and observing the response, or making a model of it and then reducing that model. In biology it is impossible to avoid reduction, even if we wished. The material and relationships of living things are so phenomenally complex that a holistic view is seen only when it suits the biota to exist as an identifiable entity such as a cell, a plant, a nest, or Gaia. Certainly, the entities themselves can be observed and classified with a minimum of invasion, but sooner or later curiosity will drive an urge to discover what the entities are made of and how they work. In any case, the idea that mere observation is neutral is itself an illusion. Someone once said that the reason the Universe is running down is that God is always observing it and hence reducing it. Be this as it may, there is little doubt that a nature reserve, a wildlife park, or an ecosystem is reduced in proportion to the amount of time that we and our children perturb the wildlife by watching them.

In *The Self-Organizing Universe,* Erich Jantsch made a strong argument for the omnipresence of a self-organizing tendency; so that life, instead of being a chance event, was an inevitable consequence. Jantsch based his thoughts on the theories of those pioneers of what might be called the "thermodynamics of the unsteady state"—Max Eigen, Ilya Prigogine, Humberto Maturana, Francisco Varela, and their successors. As scientific evidence accumulates and theories are developed in this recondite topic, it may become possible to encompass the metaphor of a living Universe. The intuition of God could be rationalized; something of God could become as familiar as Gaia.

For the present, my belief in God rests at the stage of a positive agnosticism. I am too deeply committed to science for undiluted faith; equally unacceptable to me spiritually is the materialist world of undiluted fact. Art and science seem interconnected with each other and with religion, and to be mutually enlarging. That Gaia can be both spiritual and scientific is, for me, deeply satisfying. From letters and conversations I have learnt that a feeling for the organism, the Earth, has survived and that many feel a need to include those old faiths in their system of belief, both for themselves and because they feel that Earth of which they are a part is under threat. In no way do I see Gaia as a sentient being, a surrogate God. To me Gaia is alive and part of the ineffable Universe and I am a part of her.

The philosopher Gregory Bateson expressed this agnosticism in his own special way:

> The individual mind is immanent but not only in the body. It is immanent also in pathways and messages outside the body; and there is a larger mind of which the individual mind is only a sub-system. This larger mind is comparable to God and is perhaps what some people mean by God, but it is still immanent in the total interconnected social systems and planetary ecology.

As a scientist I believe that Nature is objective but also recognize that Nature is not predetermined. The famous uncertainty principle that the physicist Werner Heisenberg discovered was the first crack in the crystalline structure of determinism. Now chaos is revealed to have an orderly mathematical prescription. This new theoretical understanding enlightens the practice of weather forecasting. Previously it was believed, as the French physicist Laplace had stated, that given enough knowledge (and, in this age, computer power) anything could be predicted. It was a thrill to discover that there was real, honest chaos decently spread around the Universe and to begin to understand why it is impossible in this world ever to predict if it will be raining at some specific place or time. True chaos is there as the counterpart of order. Determinism is reduced to a collection of fragments, like jewels that have fallen on the surface of a bowl of pitch.

Science has its fashions, and one thing guaranteed to stir interest and start a new fashion is the exploration of a pathology. Health is far less interesting than disease. I well recall as a schoolboy visiting the Museum of the London School of Hygiene and Tropical Medicine where there were on display life-sized models of subjects stricken by tropical illnesses. Although less well crafted, they were so strange and horrible as to make tame the professional horrors of Madame Tussaud's waxworks. The sight of full-sized models of the victims of elephantiasis or leprosy and the imagination of their suffering made bearable the adolescent agonies of a schoolboy. Contemporary science is similarly fascinated by pathologies of a mathematical kind. Theoretical ecology, as we have already discussed, is more concerned with sick than with healthy ecosystems. The vagaries of weather are more interesting than the long-term stability of climate. Continuous creation never had a chance in face of the ultimate pathology of the Big Bang.

Interest in the pathologies of science has a curious link with religion. Mathematicians and physicists are, without seeming aware of it, into demonology. They are found investigating "catastrophe theory" or "strange attractors." They then seek from their colleagues in other sciences examples of pathologies that match their curious models. Perhaps I should explain that in mathematics, an attractor is a stable equilibrium state, such as a point at the bottom of a smooth bowl where a ball will always come to rest. Attractors can be lines, planes, or solids as well as points, and are the places where systems tend to settle down to rest. Strange attractors are chaotic regions of fractional dimensions that act like black holes, drawing the solutions of equations to their unknown and singular domains. Phenomena of the natural world—such as weather, disease, and ecosystem failures—are characterized by the presence of these strange attractors in the clockwork of their mathematics, lurking like time bombs as harbingers of instability, cyclical fluctuations, and just plain chaos.

The remarkable thing about real and healthy living organisms is their apparent ability to control or limit these destabilizing influences. It seems that the world of dissipating structures, threatened by catastrophe and parasitized by strange attractors, is the foreworld of life and of Gaia and the underworld that still exists. The tightly coupled evolution of the physical environment and the autopoietic entities of pre-life led to a new order of stability; the state associated with Gaia and with all forms of healthy life. Life and Gaia are to all intents immortal, even though composed of entities that at least include dissipative structures. I find a curious resemblance between the strange attractors and other denizens of the imaginary world of mathe-matical constructs and the demons of older religious belief. A parallel that goes deep and includes an association with sickness not health, famine not plenty, storm not calm. A saint of this fascinating branch of mathematics is the Frenchman, Benoît Mandelbrot. From his expressions in fractional dimensions it is possible to produce graphic illustrations of all manner of natural scenes: coastlines, mountain ranges, trees, and clouds, all star-tlingly realistic. But when Mandelbrot's scientific art is applied to strange attractors we see, in graphic form, the vividly colored image of a demon or a dragon.

Gaia theory may seem to be dull in comparison with these exotica. A thing, like health, to be taken for granted except when it fails. This may be why so few scientists and theologians are interested in it; they prefer the exploration of the Universe, or of the origins of life, to the exploration of the natural world that surrounds them. I find it difficult to explain to my colleagues why I prefer to live and work alone in the depths of the country. They think that I must be missing all the excitement of explora-tion. I prefer a life with Gaia here and now, and to look back only to that part of her history which is knowable, not to what might have been before she came into being. A friend has asked why, if this is so, I chose to spend so much of this book on the history of the Earth. I find it easiest to explain my reasons for this apparent inconsistency in a fable.

Imagine an island set in a warm blue sea with sandy beaches. The lush forest in the foreground gives way to small rocky mountain peaks as sharp and clear as a line drawing on the distant horizon. There is no sign of habitation, human or other. What at first sight looks like a village of white stone houses turns out, on closer inspection, to be a chalk outcrop, laser bright in the sunlight. Something looks odd, though; you blink, for the light is very bright, and look again. It is not an illusion, the trees are not green, they are a dark shade of blue.

The island in view is somewhere on Earth 500 million years from now. The exact details are unpredictable and unimportant to this travel tale, but we can say it is hotter than any seaside place on Earth today, with a sea temperature near 30°C; it often reaches 60°C in the desert inland. There is little or no carbon dioxide in the air, but otherwise it is much the same as now, with just the right amount of oxygen for breathing but not so much as to make fires uncontrollable. There has been a major punctuation, and the dominant life forms on the land surface are of a structure no botanist or zoologist of our time would recognize.

In a small meadow near the shore, a group of philosophers is gathered for one of those civilized meetings hosted by a scientific society. A symposium that leaves ample time for swimming and walking and just talking idly. A participant has a theory that their form of life, so unlike that of many of the organisms in the sea and of the microorganisms, did not just evolve but was made artificially by a sentient life form living in the remote geological past. She bases her argument on the nature of the nervous system of the philosophers and of land animals generally. It operates by direct electrical conduction along organic polymer strands, whereas that of the ocean life operates by ionic conduction within elongated cells (which we, of course, would recognize as nerves). The brains of the philosophers operate by semiconduction, in contrast to the chemically polarized systems of the sea organisms. In this new form of life, males do not exist as mobile sentient organisms, merely as a vegetative form that supplies the necessary separate pathway for genetic information so that recombination can reduce the expression of error. Marriage is still a lifelong relationship, but with males rooted in the soil like plants, it is more one of that between a loving gardener and the flowers. Our philosopher argues that such a system could never have originated by chance but must have been manufactured at some time in the past. Not surprisingly, her theory is not well received. Not only is it outside the paradigm of the science of those times, but the theologians and mytho-poets find the notion repugnant to their view of a single, sponta-neous origin of a living planet. To bring back the Creationist heresy is unacceptable.

These occupants of a future Atlantis have no need for speech or writing. The possession of an electronic nervous system makes speech redundant; they are able to use radio frequencies to communicate directly a wide range of images and ideas. In spite of these advantages and their superior wisdom, they are, like the whales of today, neither mechanically adept nor interested in mechanisms. This being so, the very idea of making anything as intricate as a brain or nervous system as an artifact is beyond their understanding, and therefore, in their minds, beyond the capabilities of a past life form.

The point of the fable is to argue that it is not necessary to know the intricate details of the origin of life itself to understand the evolution of Gaia and of ourselves. In a similar way, the contemplation of those other remote places before and after life, Heaven and Hell, may be irrelevant to the discovery of a seemly way of life. We may well have been assisted by the nature of the Universe to cheat chaos and evolve spontaneously, on some Hadean shore, into our ancestral form of life. It seems unlikely that we come from a life form planted here by visitors from elsewhere; or even arrived clinging to some piece of come-tary debris from outer space. I like to think that Darwin dismissed enquiries about the origins of life not merely because the informa-tion available in his time was so sparse that the search for life's origin would have had to remain speculative, but, more cogently, because he recognized that it was not necessary to know the details of the origin of life to formulate the evolution of the species by natural selection. This is what I mean by the concept of Gaia being manageable.

The belief that the Earth is alive and to be revered is still held in such remote places as the west of Ireland and the rural parts of some Latin countries. In these places, the shrines to the Virgin Mary seem to mean more, and to attract more loving care and attention, than does the church itself. The shrines are almost always in the open, exposed to the rain and to the sun, and surrounded by carefully tended flowers and shrubs. I cannot help but think that these country folk are worshipping something more than the Christian maiden. There is little time left to prevent the destruction of the forests of the humid tropics with consequences far-reaching both for Gaia and for humans. The country folk, who are destroying their own forests, are often Christians and venerate the Holy Virgin Mary. If their hearts and minds could be moved to see in her the embodiment of Gaia, then they might become aware that the victim of their destruction was indeed the Mother of humankind and the source of everlasting life.

When that great and good man Pope John Paul travels around the world, he, in an act of great humility and respect for the Mother or Father Land, bends down and kisses the airport tarmac. I sometimes imagine him walking those few steps beyond the dead concrete to kiss the living grass; part of our true Mother and of ourselves.

Epilog

I will not cease from Mental Fight,
Nor shall my sword sleep in my hand
Till we have built Jerusalem
In England's green and pleasant Land.
      WILLIAM BLAKE, *Milton*

In letters and conversation, people often ask, "How should we live in harmony with Gaia?" I am tempted to reply, "Why ask me? All that I have done is to see the Earth differently; that does not qualify me to prescribe a way of life for you." Indeed, after nearly twenty years of writing and thinking about Gaia, it still seems that there is no prescription for living with Gaia, only consequences. Knowing that the question about how to live with Gaia is serious and that such a reply would be discourteous, as well as unhelpful, I will try to show what living with Gaia means to me. Then, perhaps, the questioner will discover something that we share in common.

My life, as a scientist-hermit, would suit very few. Most people are gregarious and enjoy the lively chattering of human company in pubs, churches, and parties. Living alone with Nature, even as a family unit, is not for them. So let me take you on a tour around the place where we live in north Devon, and as you walk with me I will try to explain why we prefer to live as we do. Then maybe you will see your own way to live with Gaia.

Soon after Helen and I came to live at Coombe Mill we adopted a peacock and a peahen. It was a delusion of grandeur coming from recollections of stately homes where peafowl strutted sedately and displayed their amazing colored tails. Coombe Mill is, in fact, a small cottage with thick mud-and-straw walls and a slate roof, an English adobe. But we did have 14 acres of land to start with, now grown to 30 acres. With the nearest neighbor about a mile away, it is room enough to keep the noisiest birds there are. Noisy they may be, but to us their triumphant trumpet sound at mating time is fitting and seems to usher in the spring. For the rest of the year, their extensive vocabulary ranges from a gentle clucking or purring sound to cries like a donkey braying. Then there is the sharp bark of their alarm call when, all too often, wild dogs stray onto our land. Helen, the dedicated gardener and keeper of our environment, calls them mobile shrubs, and we both have enjoyed their colorful company over the years. There is only one disadvantage—their habit, either through friendliness or the expectation of snacks, of gathering on the pavement outside the door. There they leave their smelly dung. I used to curse them when I trod in it unawares or had to clean it up. But then it came to me that I was wrong and they were right. Those ecologically minded birds were doing their best to turn the dead concrete of the path back to living soil again. What better way to digest away the concrete than by the daily application of nutrients and bacteria in the shedding of their shit?

Why should we need 30 acres to live on? We are not farmers. I think the purchase of a house with so large a garden was a reaction to the changes that took place in our last village, Bowerchalke, some 130 miles to the east. In the twenty years that we lived there, we saw a living village dispossessed of its country people, and its hinterland of seemly countryside destroyed. It was a quiet rape and pillage, no savage hordes swept upon us from the downs. The destruction was by a thousand small changes over the years, until the match between our model of what the countryside should be and the reality no longer coincided. To a casual visitor the village would have looked as beautiful as ever, but with each year that passed the farms underwent metamorphosis into agribusiness factories. Fields that in the summer were Wiltshire's glory, scarlet with poppies among the grain, became a uniform green sea of weed-free barley. Meadows that once had been gardens of wildflowers were plowed and sown with a single highly productive strain of grass. When we moved, we were determined to find a place where the environment was not likely to change so drastically again. The best way to achieve this seemed to be to find a house with enough land around it to allow us to control what happened to it.

I first saw Bowerchalke in 1936 on a journey by bicycle across southern England during a summer holiday from school. Of all the places between Kent and Cornwall that I traveled, none left so lasting a memory of perfection, and I resolved there and then that one day it would be my home. I had planned my journey with the single-mindedness of a general going to war. Like him, I scrutinized ordnance maps, one inch to the mile. So detailed were these maps that they marked almost every house and tree, and finely drawn contour lines conveyed the lay of the land. I spent most winter evenings imagining the places I would visit. In those days there were few cars, and fewer still traveled on the minor roads I intended to use. With the aid of the ordnance maps, I traced a path through the network of winding lanes that joined in vertices at the villages and hamlets. Each county had its own style of architecture and its own accent. My journey was about 500 miles long and lasted for two weeks. The scale of life in England then made such a journey seem as much an expedition as does a trip to Australia now. It was not that we were diminished; it was the slower and more human pace of travel which enlarged the world.

As a novice scientist I was interested in things like wild plants, especially the poisonous ones like henbane, aconite, and deadly nightshade. I experimented once by chewing a fraction of a leaf of one of them and learnt the hard way the discomfort of atropine poisoning. Fossils too had a fascination, and the coastline of Dorset and Devon, where they lie as pebbles on the beaches, was part of my itinerary. I was led to Bowerchalke by the strange names of the Wiltshire and Dorset villages. I had to see what Plush, Folly, and Piddletrenthide looked like. I had to discover what Sydling St Nicholas was, and hear the sonorous sounding Whitchurch Canonicorum. To reach these villages, my map showed that I had to follow the Ebble Valley that led through Bowerchalke in a gentle rising slope to the high downs of Dorset. The only tight-packed contour lines, marking a steep hill, were at the head of the valley just beyond Bowerchalke, an ideal road for a traveling by bicycle.

I can still remember passing up the road from Broadchalke, with the watercress beds on my left, and rounding a corner to see before me the small thatched village of Bowerchalke, the stage of an amphitheatre of green and shrubby downland hills. I arrived there at about four on a sunny Sunday afternoon in July. I was thirsty but, unusually, there were no signs outside the cottages offering teas. In those days walkers and cyclists were common enough to make it worth the while of villagers to sell refreshments. So remote was this region, and so few the travelers, that such efforts would have brought a poor return. I asked a man walking if there was anyone who would supply my needs, and he said, "Why, yes, Mrs. Gulliver in the white cottage over there sometimes will make you a tea"; and she did. It was the memory of the quiet tranquility of Bowerchalke then, when the countryside and the people merged in a natural seemliness, free from any taint of the city, that lingered in my mind and brought me back some twenty years later to make it our family home.

The recent act of destruction of the English countryside is a vandalism almost without parallel in modern history. Blake saw the threat of those dark satanic mills a century ago, but he never knew that one day they would spread until the whole of England was a factory floor. Humans and Nature had evolved together to form a system that sustained a rich diversity of species; something that stirred poets and even Darwin, who wrote about the mystery of the "tangled bank." It was so familiar, so taken for granted, that we never noticed its going until it was gone. Had anyone proposed building a new road through the close of Salisbury Cathedral the reaction would have been immediate. But farmers were paid by the Ministry of Agriculture to emulate the prairies, those man-made deserts in which nothing grows but grain, and nothing lives but farmers and their livestock. The yearly tromp of vast and heavy machines and the generous spraying of herbicides and pesticides ensured that all but a few resistant plant and insect species were eliminated. The older-style farmers could not stomach it, and left the land to young agricultural college graduates working as managers for city institutions. One old farmer said to me, "I didn't do farming to be a mechanic in a factory." But it was wonderfully efficient, and soon England was producing far more food than could be eaten.

The destruction still goes on. Even here in Devon, the hedge-rows and small copses still fall to the chain saws and diggers. Rachel Carson was right in her gloomy prediction of a silent spring, but it has come about not simply by pesticide poisoning, as she imagined, but by the attack on all fronts of the farmers enemies, "weeds, pests, and vermin." Birds need a place to nest, and where better than the hedges, those marvelous linear forests that once divided our fields. Government, on the advice of negligent civil servants, paid handsome subsidies to farmers to root out the hedgerows, until the wildlife was destroyed, just as effectively as if the land had been sprayed with pesticide. The environmentalists, who should have seen what was happening and protested before it was too late, were much too busy fighting urban battles, or demonstrating outside the nuclear power stations. Their battle, whatever was claimed otherwise, was more against authority, represented by the monolithic electricity supply board, than for saving the countryside. They sometimes noticed poisonous sprays, for they were the products of the hated multinational chemical industries. But few were the friends of the soil who protested the agribusiness farms, or noticed the mechanized army of diggers and cutters working to make the landscape sterile for next year's planting of grain. There is no excuse for their neglect. Marion Shoard, in her moving and well-publicized book, *The Theft of the Countryside,* said all that I have said and much more.

To those who see the world in terms of a conflict between human societies and groupings for power, my personal view of the changing landscape must seem obsessional and irrelevant. They also are the vast majority everywhere, whether in the cozy comfort of air-conditioned suburban homes of the First World, or in the squalor of a Bidonville.

Who was most to blame for the destruction? Without doubt it was the scientists and agronomists who worked to make farming efficient. The experience of near starvation in the Second World War was a powerful stimulant to make Britain self-sustaining in food. Their intentions were good, it was just that they could not foresee the consequences. I know, because I was a small part of it. In my role of inventor, I helped friends and colleagues at the Grassland Research Institute near Stratford-upon-Avon in the 1940s. They were intent on improving the output of food from the small-scale English farms. I recall their sermons to young farmers on the inefficiency of hedgerows that hindered the free movement of machinery around a field; on the waste of meadows left as permanent pasture compared with a good crop of Italian rye grass grown as a monoculture. We never dreamt that the message would be so well heard that the government would be persuaded to pass the legislation that led to the removal of hedges and to the nurturing of agribusiness. Nor did we have the imagination to see that most young farmers share, with young males everywhere, a delight in mechanical toys. We, and through us the government, were giving them the money to buy, and the license to use, some of the most dangerously destructive weapons ever used. Weapons to fight the farmer's enemies, which were all life other than crops, livestock, hired help, and the farmer's family.

Should anyone think that I have got it wrong, that this was another example of heartless exploitation done by a government of capitalist nominees for the profit of a few multinationals, I would remind them that it started in the late 1940s during the period of the post-war Labor government, an administration secure in power, confident, and committed to its socialism. The destruction of the countryside was independent of politics; it was carried through by good intentions aided by the tendency of civil servants to apply positive feedback by subsidies, or a negative one through taxes. Farmers work on very small margins. They may own land worth up to a million pounds, but their returns may be very small compared with the returns from simple investment. A minuscule subsidy can turn a slight loss into a comfortable profit. The countryside has vanished from most of England, and what little remains here in the West Country is passing away because the government continues to pay farmers a subsidy which is just enough to make it worth their while to act as destroyers rather than as gardeners. The small subsidy to remove hedgerows has led to the loss of over 100,000 miles of them in the past few decades. An equally small subsidy would put them back again, although it would be generations before they served once more as the linear ecosystems and artistic landscape features of the countryside.

So what should we do instead? My vision of a future England would be like Blake's: to build Jerusalem on this green and pleasant land. It would involve the return to small, densely populated cities, never so big that the countryside was further than a walk or a bus ride away. At least one-third of the land should revert to natural woodland and heath, what farmers now call derelict land. Some land would be open to people for recreation; but one-sixth, at least, should be "derelict," private to wildlife only. Farming would be a mixture of intensive production where it was fit so to be, and small unsubsidized farms for those with the vocation of living in harmony with the land. In recent years, the overproduction of food by immoderate farming in the European Economic Community, including England, has been so vast that events have made my vision the basis of a practical plan for countryside management.

In their humorless despair, I have sometimes heard Greens parody Sir John Betjeman's verse, written near the beginning of the Second World War:

Come, friendly bombs, and fall on Slough,
To get it ready for the plough.
The cabbages are coming now;
The Earth exhales.

with

Come friendly nukes and strike them down
And blast their ever spreading town . . .

Even for the desperate, such an evil catharsis is not needed. Left to herself, Gaia will relax again into another long ice age. We forget that the temperate Northern Hemisphere, the home of the rich First World, now enjoys a brief summer between long, long periods of winter that last for a hundred thousand years. Even the nukes would not so devastate the land; nor would a "nuclear winter," if it could happen at all, last long enough to return the land to its normal frozen state. The natural state here in Devon has been, for most of the past million years, a permanent arctic winter. Even though close to the ocean, it was still as bitterly cold and barren as is Bear Island in the Arctic Ocean now. A mere 50 miles to the north or east of Coombe Mill were the great permanent glaciers of the "ice ages." These bulldozer blades of ice scraped off every vestige of surface life that had flowered in the brief interglacials like now.

So why should I fret over the destruction of a countryside that is, at most, only a few thousand years old and soon to vanish again? I do so because the English countryside was a great work of art; as much a sacrament as the cathedrals, music, and poetry. It has not all gone yet, and I ask, is there no one prepared to let it survive long enough to illustrate a gentle relationship between humans and the land, a living example of how one small group of humans, for a brief spell, did it right?

The little that is left of old England is still under threat. The donnish guardians of the landscape seem unaware of its existence. They see the countryside through romantic notions of scenic beauty. In my part of Devon they look only at the tundra of Dartmoor, and see it as something of inestimable value to be preserved at all costs. Tundra—the waterlogged bog, too wet and too cold for trees to grow—is a common place where the polar and temperate zones merge, a memory of what this region was in the last glaciation. In great contrast, the same guardians regard the land to the north of Dartmoor, with its small low-efficiency farms, rich wildlife, and village communities that have changed little since the Domesday Book, as of no account and expendable, a fit place for new schemes, such as a reservoir, a new road, or an industrial site.

I often think that those city planners who act so destructively have been misled by that great novelist Thomas Hardy. His writing deeply influenced my city-born and city-bred mother, a woman who easily saw the countryside through Hardy's distorting spectacles. My father, though, was born in Hardy's Wessex, and he showed me how very different was the reality. Hardy, for all the brilliance of his characterization, did not understand the countryside and used it merely as a background to act out his own tragic view of the human condition.

The England I knew as a child and a young man was breathtakingly beautiful, hedgerows and small copses were abundant, and small streams and rivers teemed with fish and fed the otters. It inspired generations of poets to make coherent the feelings we could not ourselves express. Yet that landscape of England was no natural ecosystem; it was a nation-sized garden, wonderfully and carefully tended. The degraded agricultural monocultures of today—with their filthy batteries for cattle and poultry, their ugly sheet-metal buildings, and roaring, stinking machinery—have made the countryside seem to be a part of Blake's dark satanic mills. I know it seems that way because I knew it as it was. Visitors come to Coombe Mill from the cities and abroad, and eulogize over the few glories that remain. They, and the planners of the countryside, do not understand that, unless we stop the ecocide soon, Rachel Carson's gloomy prediction of a silent spring will come true, not because we have poisoned the birds with pesticides, but because we have destroyed their habitats, and they no longer have anywhere to live.

Being a typical Englishman, I did not expect "them," the establishment, to change their ways. There was nothing for it but for my family to try to do our best with the land we owned at Coombe Mill; make it a habitat and a refuge for some of the plants and animals that agribusiness is destroying. This is how we, personally, choose to live with Gaia.

There are only three of us here, but 30 acres is not much more difficult to manage than a suburban garden. A garden lawn forever needs mowing, feeding, watering, and weeding; a ceaseless labor or a cost if someone else is to do it. Ten acres of our land is grass. It is no nightmare lawn requiring the ceaseless attention of an army of gardeners; it grows as meadows rich with wildflowers and small animals. The meadows divide and form a setting for the 20 acres of planted trees. It needs only to be enjoyed and cut once a year when the grass has grown long. Local farmers are glad to come and cut it; they use the grass for fodder and pay for it. The cost of keeping 10 acres of meadow is comparable with that of a well-kept suburban garden. The trees need more attention, but not so much as to be in any way a burden for the three of us.

The River Carey divides our land into two equal parts, which posed a problem. The river runs by the house, which was once a water mill, and is about 60 feet wide. It cannot easily be crossed by wading, and we soon discovered that to reach our new land involved a five-mile walk. Bridges across the Carey are widely spaced apart. Two years ago, we decided to build a bridge so that we could more easily tend the 10,000 trees that were newly planted on the west bank of the Carey. As metaphors go, building a bridge has almost become a cliché. But just try building a bridge in real life; it is amazing to experience, person- ally, the power of reducing a metaphor to practice.

As you will have gathered, we are solitary people and don't much mix with our neighbors. Yet in this part of west Devon we were welcomed as soon as we came and have experienced more spontaneous kindness than in any other place that we have lived. Helen and I and our son John are in various ways physically handicapped so that we add up to make one able-bodied person, not enough to run a place as large as this. It would not have flourished had it not been for the unstinting care and generous help of our friends from the village, Keith and Margaret Sargent. Our home and the buildings that go to make up the rest of this place are of mud and straw, with slate roofs. They would never have survived the winter storms but for the skillful repairs of our other village friends, and former occupants of Coombe Mill, Ernie and Bill Orchard. But it was not until we started to plan our bridge that we experienced the full vigor of the community in which we are immersed.

When these friends knew what was in our minds, the bridge began to form—first in the imagination as an exciting project, and then more solidly as the plans were drawn and the materials gathered. They had the skills needed to do with élan and enjoy-ment a challenging task, one that arose from no more than a passing personal thought. The project showed, in a Gaian way, how a thought became an act that brought personal and then local benefit.

Our bridge is made of steel; it was built by a blacksmith, Gilbert Rendall, and is in every way a mechanical construction. I am never quite comfortable with things mechanical. I well recall a conversation with my friends Stuart Brand, editor of *CoEvolution Quarterly,* and Gary Snyder, the poet. They were shocked and indignant when I said, "Chain saws are an invention more evil than the hydrogen bomb." To me a chain saw was something that cut down in minutes a tree that had taken a hundred years to grow. It was the means of destroying the tropical forests. To Gary Snyder it was a benign gardening tool with which he could carefully, like a surgeon, remove the scars of years of bad husbandry in his forests. It is not what you do but the way it is done; the more powerful the tool, the harder it is to use it right.

How, you may ask, do these rambling thoughts tell us about how to live with Gaia? I would reply that as a metaphor, Gaia emphasizes most the significance of the individual organism. It is always from the action of individuals that powerful local, regional, and global systems evolve. When the activity of an organism favors the environment as well as the organism itself, then its spread will be assisted; eventually the organism and the environmental change associated with it will become global in extent. The reverse is also true, and any species that adversely affects the environment is doomed; but life goes on. Does this apply to humans now? Are we doomed by our destruction of the natural world? Gaia is not purposefully antihuman, but so long as we continue to change the global environment against her preferences, we encourage our replacement with a more environmentally seemly species.

It all depends on you and me. If we see the world as a living organism of which we are a part—not the owner, nor the tenant; not even a passenger—we could have a long time ahead of us and our species might survive for its "allotted span." It is up to us to act personally in a way that is constructive. The present frenzy of agriculture and forestry is a global ecocide as foolish as it would be to act on the notion that our brains are supreme and the cells of other organs expendable. Would we drill wells through our skins to take the blood for its nutrients? If living with Gaia is a personal responsibility, how should we do it? Each of us will have a personal solution to the problem. There must be many simpler ways of living with Gaia than the one we have chosen at Coombe Mill. I find it useful to think of things that are harmless in moderation but malign in excess. For me these are the three deadly Cs: cars, cattle, and chain saws. For example, you could eat less beef. If you do this, and if the clinicians are right, your health might improve and at the same time you would ease the pressures to turn the forests of the humid tropics into absurdly wasteful beef farms.

Gaia theory arose from a detached, extraterrestrial view of the Earth, too distant to be much concerned with humans. Strangely, the view is not inconsistent with the human values of kindness and compassion; indeed it helps us to reject sentimentality about pain and death, and accept mortality, for us as well as for our species. With such a view in mind, Helen and

I wish our eight grandchildren to inherit a healthy planet. In some ways, the worst fate that we can imagine for them is to become immortal through medical science—to be condemned to live on a geriatric planet, with the unending and overwhelming task of forever keeping it and themselves alive for our kind of life. Death and decay are certain, but they seem a small price to pay for the possession, even briefly, of life as an individual. The second law of thermodynamics points the only way the Universe can run—down, to a heat death. The pessimists are those who would use a flashlight to see their way in the dark and expect the battery to last forever. Better to live as Edna St. Vincent Millay advised:

> My candle burns at both ends;
>   It will not last the night;
> But, ah, my foes, and, oh, my friends—
>   It gives a lovely light.

References

Hutton, James. 1788. "Theory of the Earth; or an investigation of the laws observable in the composition, dissolution, and restoration of land upon the globe." Roy. Soc. Edinburgh, Tr., 1, 209–304.

Doolittle, W. F. 1981. "Is Nature Really Motherly?" *CoEvolution Quarterly* 29, 58–63.

Holland, H. D. 1984. *The Chemical Evolution of the Atmosphere and the Oceans.* Princeton, N.J.: Princeton University Press, 539.

Lovelock, J. E. 1972. "Gaia as Seen through the Atmosphere." *Atmospheric Environment* 6, 579–80.

Lovelock, J. E.; Maggs, R. J.; and Rasmussen, R. A. 1972. "Atmospheric Dimethyl Sulphide and the Natural Sulphur Cycle." *Nature* 237, 452–53.

Margulis, L., and Lovelock, J. E. 1974. "Biological Modulation of the Earth's Atmosphere." *Icarus* 21, 471–89.

Charlson, R. J.; Lovelock, J. E.; Andreae, M. O.; and Warren, S. J. 1987. "Ocean Phytoplankton, Atmospheric Sulfur, Cloud Albedo and Climate." *Nature* 326, 655–61.

Whitfield, M., and Turner, David R. 1987. "The Role of Particles in Regulating the Composition of Seawater." In *Aquatic Surface Chemistry*, ed. Werner Stumm. Chichester, Eng.: Wiley.

Further Reading

CHAPTER 1

For an alternative view of the work at JPL, what would be better to read than *The Search for Life on Mars,* by Henry Cooper (New York: Holt, Rinehart and Winston, 1976)?

CHAPTER 2

A beautiful and entirely comprehensible book about entropy is P. W. Atkins's *The Second Law* (New York and London: Freeman, 1986).

The classic account of the problem of defining life is *What Is Life?,* by Erwin Schrödinger, written in Dublin during the Second World War (Cambridge: Cambridge University Press, 1944).

The lightest of Ilya Prigogine's books on the difficult subject of the thermodynamics of the unsteady state is *From Being to Becoming* (San Francisco: Freeman, 1980).

For a straightforward account of the evolution of the Earth from a geologist's viewpoint, there is no better book than *The Chemical Evolution of the Atmosphere and Oceans,* by H. D. Holland (Princeton, N.J.: Princeton University Press, 1984).

A similar book on climatology is *The Coevolution of Climate and Life,* by Stephen Schneider and Randi Londer (San Francisco: Sierra Club Books, 1984).

CHAPTER 3

Those interested in geophysiological theory should read Alfred Lotka's classic book, *Elements of Physical Biology* (Baltimore: Williams and Wilkins, 1925).

Autopoiesis, the organization of living things, and many other concepts helpful for understanding life as a process, are described in *The Tree of Knowledge,* by Humberto R. Maturana and Francisco J. Varela (Boston: New Science Library, 1987).

CHAPTER 4

For those interested in the synthesis of elements within stars and about the life of stars in general, there is a splendid account of these awesome events in Sir Fred Hoyle's *Astronomy and Cosmology* (San Francisco: Freeman, 1975).

In *Early Life,* Lynn Margulis provides a beautiful and clearly written account of the known and the conjectured of the obscure period before and after life began, including a picture of the evolution of nascent life (Boston: Science Books International, 1982).

If you are interested in the beginnings, then read *Origins of Life,* by Freeman Dyson (Cambridge: Cambridge University Press, 1986) and *Origins: A Skeptic's Guide to the Creation of Life on Earth,* by Robert Shapiro (New York: Summit, 1986).

A rare geological text that restores soul to the misty Archean world is E. G. Nisbet's book, *The Young Earth* (London: Allen and Unwin, 1986).

The work of some of the pioneers of the new field of biomineralization is recorded in *Biomineralization and Biological Medical Accumulation,* by P. Westbroek and E. W. de Jong (Dordrecht: Reidel, 1982).

For those interested in the evolution of eukaryotic cells, there is a detailed account in Lynn Margulis's book *Symbiosis in Cell Evolution* (San Francisco: Freeman, 1981).

A splendid account of the four eons of evolution from our microbial ancestors is in Lynn Margulis and Dorion Sagan's book *Microcosmos* (New York: Simon and Schuster, 1986).

CHAPTER 6

A fair amount of Gaia theory has come from evidence about the atmosphere and atmospheric chemistry. A book that summarizes the evidence of this subject in a readable way is *Chemistry of Atmospheres,* by Richard P. Wayne (Oxford: Oxford University Press, 1985).

For a professional account drawn from conventional wisdom, *The Planets and Their Atmospheres* by John S. Lewis and Ronald G. Prinn is to be recommended as an antidote to Gaia (Orlando, Fla.: Academic Press, 1984).

CHAPTER 7

The proceedings of the meeting in Brazil mentioned in the chapter are now published as a book, *The Geophysiology of Amazonia,* edited by Robert E. Dickinson (New York: Wiley, 1987).

An account of the battlefield scenes of the chlorofluorocarbon conflict is to be found in *The Ozone War,* by Lydia Dotto and Harold Schiff (New York: Doubleday, 1978).

Rachel Carson stands, like Marx, as the major influence behind a revolution, this time in environmental thought and action. Her seminal book, *Silent Spring* (Boston: Houghton Mifflin, 1962), must be included in any further reading of the topics related to this chapter.

Environmental affairs are in the realm of politics, and for a wise and understanding view from that perspective you should read *Climatic Change and World Affairs,* by Sir Crispin Tickell (Lanham, Md.: University Press of America, 1986).

An outstanding figure among environmental scientists is Paul R. Erlich, his book with Anne H. Erlich, *Population Resources Environment,* is essential reading to capture the heart and mind of the ecology movement (New York: Freeman, 1972).

A contemporary view of environmental problems is provided in *Sustainable Development of the Biosphere,* edited by William C. Clark and R. E. Munn (Cambridge: Cambridge University Press, 1986).

CHAPTER 8

If you really want to know what the surface of Mars looks like, then there is no better source than the descriptive writing and photographs in Michael Carr's beautiful book, *The Surface of Mars* (New Haven and London: Yale University Press, 1981).

CHAPTER 9

A theologian's view of Gaia is expressed in Hugh Montefiore's *The Probability of God* (London: SCM Press, 1985).

An unusual and very readable book is *God and Human Suffering,* by Douglas John Hall. Although it is not directly concerned with Gaia, I found it to be helpful and moving while revising this chapter (Minneapolis: Augsburg, 1986).

For me the most important book to connect with this chapter is *Angels Fear: Towards an Epistemology of the Sacred,* by Gregory Bateson and Mary Catherine Bateson (New York: Macmillan, 1987).

For an understanding of scientists' views of the Universe, perhaps the best summary is in *The Self-Organizing Universe,* by Erich Jantsch (Oxford: Pergammon, 1980).

A subject often linked with Gaia, but which is in fact very different, is *The Anthropic Cosmological Principle,* by John D. Barrow and Frank J. Tipler (Oxford: Oxford University Press, 1986).

GENERAL

For those who find the topic of Gaia entertaining, probably no one has written with more feeling than Lewis Thomas in his many books, in particular *The Lives of a Cell* (New York: Viking Press, 1975), and *The Youngest Science* (New York: Viking Press, 1983).

No guide to the world would be complete without an atlas, and the most appropriate would be *Gaia: An Atlas of Planet Management,* edited by Norman Myers (New York: Doubleday, 1984).

The evolution of the Earth from a geologist's viewpoint is clearly expressed and beautifully illustrated by Frank Press and Raymond Siever in *Earth* (San Francisco: Freeman, 1982).

For a physicist's view of the evolution of the cosmos and life, see Eric Chaisson's book, *The Life Era* (New York: Atlantic Monthly Press, 1987).

And for a mainstream ecologist's view of the Earth written at the same time but in great contrast to the views expressed in the first Gaia book, I strongly recommend Paul Colinvaux's book, *Why Big Fierce Animals Are Rare* (Princeton, N.J.: Princeton University Press, 1978).

The world of the scientists who have participated in the discoveries recorded in this book is captured in *Planet Earth,* by Jonathan Weiner (New York: Bantam Books, 1986).

Index

About the Author

Born in 1919, James Lovelock was educated at the University of London and
Manchester University and holds a Ph.D.
in medicine. In the United States, he has
taught at Yale, the Baylor University College of Medicine, and, as a Rockefeller
Fellow, at Harvard. An independent scientist, Lovelock works out of a barn-
turned-laboratory at Coombe Mill in
Cornwall, England. He is president of the
Marine Biology Association, a fellow of
the Royal Society, London, and author of
Gaia: A New Look at Life on Earth.